Fuels and Petrochemicals Division 2016

Core Programming Area at the 2016 AIChE Spring Meeting and 12th Global Congress on Process Safety

Houston, Texas, USA
10-14 April 2016

ISBN: 978-1-5108-2494-2

Fuels and Petrochemicals Division 2016

Core Programming Area at the 2016 AIChE Spring Meeting and 12th Global Congress on Process Safety

Houston, Texas, USA
10-14 April 2016

AIChE
The Global Home of Chemical Engineers

Printed from e-media with permission by:

Curran Associates, Inc.
57 Morehouse Lane
Red Hook, NY 12571

Printed by Curran Associates, Inc. (2016)

AIChE
120 Wall Street, FL 23
New York, NY 10005-4020

Phone: (800) 242-4363
Fax: (203) 775-5177

www.aiche.org

Additional copies of this publication are available from:

Curran Associates, Inc.
57 Morehouse Lane
Red Hook, NY 12571 USA
Phone: 845-758-0400
Fax: 845-758-2633
Email: curran@proceedings.com
Web: www.proceedings.com

TABLE OF CONTENTS

OPERATING AND DESIGN TECHIQUES TO INCREASE PETROLEUM REFINERY PROPYLENE CAPACITY

Paper Number 444873

Christina J. Conrad, Ben F. Ketter and Dominic M. Varraveto
Burns & McDonnell, Kansas City MO

Prepared for Presentation at
American Institute of Chemical Engineers
2016 Spring Meeting
11th Global Congress on Process Safety
Houston, Texas
April 10-14, 2016

UNPUBLISHED

Abstract

Increasing worldwide petrochemical demand and recent developing preference for processing abundant lighter, lower molecular weight feedstock in new, as well as existing, thermal cracking plants have combined to create a long term divergence between propylene supply and demand. In addition to production from cracking plants, propylene is a byproduct from petroleum refinery catalytic cracking and thermal coking units processing vacuum gas oil and vacuum residue respectively. Propylene can also be produced on purpose by catalytically dehydrogenating propane. Many new, on purpose propylene production facilities have been announced recently and several are nearing completion or are already operating. These plants tend to be located in the U.S. close to emerging, plentiful propane supply or in Asia close to propylene demand for polypropylene and derivative production. Rich propane to propylene margins prevalent when energy prices are high support the relatively high investment costs necessary for new on purpose plants. Since the second half of 2014, energy pricing and project economics have undergone a significant downward shift causing investors to rethink investment in new facilities and look at lower capital investment alternatives. This paper focuses on operating and design techniques for optimizing petroleum refinery yield, recovery and purification of propylene. Some considerations for producing propylene derivatives will also be discussed.

Introduction

Worldwide Propylene demand growth of 4% from 90 to 130 MTA over the next decade is expected to significantly outpace projected annual gasoline demand growth, which is not expected to exceed 1%.

Factors driving Propylene demand growth:

1) Wide use of Propylene as a feedstock for chemical derivatives
2) Range of derivative product uses including films, fibers and thermoplastics
3) Desirable derivative properties including light weight, flexibility and high melt point

Factors depressing gasoline demand growth especially in the US:

1) Increasing efficiency mandated by CAFÉ, Corporate Average Fuel Economy
2) Renewable content of gasoline mandated by RFS, Renewable Fuel Standard
3) Evolution of LCFS, Low Carbon Fuel Standard in California
4) Expanding use of LNG, Diesel and Electricity for transport fuel alternatives.

In addition to demand growth disparity between gasoline and Propylene, a significant shift in the sources of Propylene production sources is already underway. The use of ethane as a preferred feedstock instead of LPG or light naphtha for Steam Cracking plants to focus on ethylene production has reduced Propylene availability from this traditional source. Table 1 shows Steam Cracker Propylene yields as a function of feedstock.

Table 1 – Steam Cracker Ethylene & Propylene Yield, wt% (from High Olefins Technology Services LLC)

Feedstock	Ethylene	Propylene	P/E
Ethane	80	3	0.04
Propane	44	15	0.34
Naphtha	30	16	0.53
Gas Oil	23	15	0.65

The data in Table 2 below summarizes the shift in production source and includes projections for the expected redistribution of supply.

Table 2 - Propylene Production Trend by Source (from IHS)

	2015	2025	Delta
Steam Cracking, %	55	40-45	-18 to -27
Petroleum Refineries, %	34	30-35	-12 to 3
On Purpose PDH, MTO, CTO, %	11	25-30	127 to 173
Total Propylene, MTA	90	130	40

The production shift is expected to be predominantly from Steam Cracking to On Purpose sources with the percentage from Petroleum Refineries remaining nearly constant. This constant percentage share of Propylene production actually represents an underlying expected 4% annual growth rate for Propylene from Petroleum Refineries. Another key consideration is that On Purpose production requires substantial investment for construction of new facilities. In contrast, a significant portion of incremental Propylene production from Petroleum Refineries can be obtained from minimal to moderate investment in existing facilities. This is a differentiator especially as energy pricing has dropped relative to project capital costs.

The main source of Petroleum Refinery Propylene is the FCCU, Fluid Catalytic Cracking Unit. Another refinery source is the DCU, Delayed Coker Unit. However, this discussion will focus on the FCCU due to the potential for exploiting the inherent FCCU operating flexibility to achieve increased Propylene production with minimal investment cost. The focus on the operation and design features for a FCCU as an incremental Propylene source will be in three sections:

1) Production in the FCCU Reactor/Regenerator
2) Recovery in the FCCU Main Fractionator and Gas Plant
3) Purification in a PP, Propylene-Propane, Splitter

A fourth section will look at considerations for production of Propylene derivatives by a Petroleum Refinery.

FCCU Propylene Production

Historically, typical Propylene yield is 3 to 5 wt% from an FCCU operated to maximize gasoline. More recently FCCU operation has shifted to increase production of LCO, Light Cycle Oil, which can be treated and blended into the distillate pool. Operating in max distillate mode reduces yield of gasoline and LPG, Liquefied Petroleum Gas, including pushing Propylene to the lower end of the typical yield range. Many factors impact FCCU product yields including feed type and quality. FCCU feed characteristics including Gravity, Distillation range, Sulfur, Metals, like Nickel, Vanadium and Iron and CCR, Conradson Carbon Residue impact FCCU performance. There are several operating parameters that can be manipulated to help achieve desired yield distribution including ROT, Reactor Operating Temperature, Catalyst Activity, C/O, Catalyst to Oil Ratio, Recycle and Steam Rates and Reactor Pressure.

In general, a higher FCCU Propylene yield is favored by increasing severity or raising ROT above 1000 °F and higher catalyst activity through the use of zeolite additives like ZSM-5. Proprietary catalyst with higher selectivity for olefins including Propylene are also available. Hydrotreating FCCU feed to reduce contaminants is also beneficial for improving FCCU yields including Propylene. Directionally increasing C/O, Recycle and Steam Rates and decreasing Reactor Pressure favor Propylene yield.

Table 3 – FCCU Propylene Yields

		Propylene Yield, wt%	Delta, %
1	FCCU	3 - 5	-
2	FCCU + ZSM-5 Additives	6 - 8	63 to100
3	Selective Catalysts	10 - 15	67 to 88
4	Design Changes	Up to 25	67 to 100

In general, each step shown above has the potential to result in a near doubling of the FCCU Propylene yield. It's possible to achieve the first incremental improvement by using additives and making operational changes staying within major FCCU equipment limits in the Reactor and Regenerator sections including items like feed nozzles, riser, termination devices, cyclones, stripper and air blower. To achieve the second and third increments shown above, more significant modifications, system re-design and replacement should be expected. Some of the possible design changes include a second riser, a "downflow" riser and/or secondary catalyst bed to maximize contact time. Doubling the C/O from 6 to 12 and increasing the riser steam rate from 10 to as high as 30 wt% may be necessary. In the extreme slurry recycle or a charge heater may be needed to provide the necessary heat input for higher C/O and greater heat of reaction to crack to olefins.

FCCU Propylene Recovery

Products from the FCCU Reactor including Dry Gas, LPG, Propylene, Cat Gasoline, LCO and Slurry are recovered and separate in a Main Fractionator, MF, and Unsats or Cracked Gas Plant, GP. Table 4 shows typical FCCU Reactor gasoline and lighter product yields operating in Gasoline vs Olefin mode.

Table 4 –FCCU yield ranges

	Gasoline	High Olefin
Wt% on Fresh Feed		
Dry Gas	1.5 - 3	3 - 12
Ethylene	0.5 – 1.5	2 -7
Total LPG	16 - 22	32 - 44
Propylene	4 - 7	12 - 22
Butylenes	4 - 8	8 -14
Gasoline	47 - 53	30 - 40

When changing operating modes from maximizing Gasoline to maximizing Olefin including Propylene, capacity of the MF and GP must be able to accommodate the shift in flow from heavier to lighter products.

In the MF this means capacity of internals in the upper section must be able to process higher vapor rates (twice as much LPG) at considerably lower liquid rates (20 to 30% less gasoline). Replacing existing fractionation trays with high capacity trays or packing may be required. In the mid and bottom sections of the MF, available cycle oil and slurry pump around heat supply will decrease, which in turn could impact integrated heat supply to the GP. Additional external steam supply maybe needed. A major concern in switching to High Olefin mode is the Wet Gas Compressor, WGC. Since the total vapor load including Dry Gas and LPG more than doubles, the WGC may have to be replaced or a second unit sized similar to the existing WGC added.

In the GP, Propylene and heavier/higher boiling temperature molecules are separated from Ethane and lighter molecules including ethylene and methane in a Rectified Absorber or RA column using cat gasoline as the absorbent. The light stream from the RA goes to a Secondary Absorber, SA, that uses LCO as a sponge oil to recover Propylene and heavier from the Dry Gas product. With the FCCU in gasoline mode, the dry gas product, primarily methane with lesser amounts of ethylene and ethane, is used refinery fuel gas. With the FCCU in High Olefin mode, the ethylene content in the overhead product from the SA may be high enough to consider recovery. The bottom stream from the RA goes to a debutanizer, DB, where a Propylene – Propane product stream is recovered as the overhead product. The debutanizer overhead product is commonly referred to as Refinery Grade Propylene and with the FCCU in High Olefin mode, the Propylene content may exceed 70%, typical refinery grade. The available capacity of the RA, SA, DB and associated reboiling, condensing and pumping equipment is suspect when switching the FCCU from Gasoline to Olefin Mode. Replacement and/or repurposing of most or all of this equipment should be expected.

Propylene Purification

As noted above refinery Propylene product is a mix of propane and Propylene. Any composition around 70% Propylene and 30% propane is considered "refinery grade". The desired composition for chemical-grade propylene is 92-96% propylene. Polymer-grade is at least 99.5% propylene.

Another way to generate a propane/propylene product is to make "on-purpose" propylene via propane dehydrogenation. This can be done with a refinery propane product or a natural gas liquids, NGL fractionation propane stream. There are at least 3 major propane dehydrogenation technologies available. With the right catalyst the reaction can be 90% selective, although side reaction can lead to unwanted products.

However the mixed propane/propylene stream is created, the two components must be separated. The high value propylene has a slightly lower atmospheric boiling point (-53.7°F) than propane (-43.6°F). Ethane has a boiling point of -128.2°F. So by comparison, ethane and propane are significantly easier to separate than propane and propylene. Distillation is the preferred separation method, but the close boiling points require a very large number of stages and significant energy.

At typical "PP Splitter" unit includes pre-treating upstream of the propane/propylene splitter column, the splitter column with its heat integration, and downstream product treating. When the feed to a splitter is refinery-grade, sulfur compounds need to be removed. Any water in the feed stream also needs to be removed. This is most common where the propane/propylene feed is being stored in an underground cavern. Propylene will be the overhead product from a PP splitter. If polymer-grade propylene is the desired product, any components lighter than propylene (ethane, ethylene, etc) should be removed upstream of the PP splitter in a deethanizer (or similar) column. Depending on the economics it might be worthwhile to recover trace propylene from the deethanizer overheads using additional refrigeration.

Depending on the desired purity of the propylene product, the PP Splitter column may have 200 to 300 stages. The reflux rate and reboiler return rates may be 8-12 times the feed rate. This leads to a very tall column with a large diameter and very high energy demands. Conventional utilities like hot oil and refrigeration can be used for reboiling and condensing, but smart designs using heat integration and heat pumps can reduce the energy requirement. Since the boiling points of the overhead stream and the bottom stream are so similar, the overhead product can be compressed and used to reboil the bottoms. This is the same concept as a heat pump and can be configured as either a closed loop or an open loop.

An interesting note for construction, high PP Splitter feed rates may make the column impossible to fabricate in one section. For these situations, the column can be split into two separate columns – an "upper" section and a "lower" section with a large vapor line and a large liquid pump to move the streams between the sections. Another option is to fabricate the column in two pieces and weld those pieces together in the field.

Even with such similar boiling points, there are still a few trace components that are intermediate boilers. The most common buildup of intermediate boilers inside a PP Splitter is methyl acetylene (propyne), which tends to accumulate near the bottom of the column.

After the overhead propylene product has been condensed, additional treating is usually required to remove arsenic compounds. Depending on the feed, the bottoms propane product may not have the desired commercial purity. Any heavy components in the propane (butane, isobutene) maybe be removed in a small debutanizer to achieve higher purity.

In summary, building a propane/propylene splitter usually involves major capitol and significant design work in the early phases of engineering. Due to the high capital costs, either the price or the quantity of propylene product must be high. For feed streams from refineries without changes to the FCC, this may mean combining feeds from multiple units or even multiple refineries. For feed streams from on-purpose propylene production or from a modified FCCU there is more likely to be sufficient flow to build dedicated PP Splitters.

The geographical location of a PP Splitter unit is important. Proximity to either the feed source(s), the propylene product destination, or access to a pipeline can dictate the better location. Building a PP Splitter unit at an existing facility may allow for smarter heat integration or utilization/expansion of existing utility systems, as long as there is room for construction and lay-down for a very tall, and possible wide, column. At atmospheric temperatures polymer-grade propylene does not have a tendency to auto-polymerize, so the product can be transported via pipeline. If polymer-grade propylene is stored in a cavern, it will need to be dried again downstream of the cavern.

Propylene Derivatives

Polypropylene: an Overview

Figure 1 (from IHS)

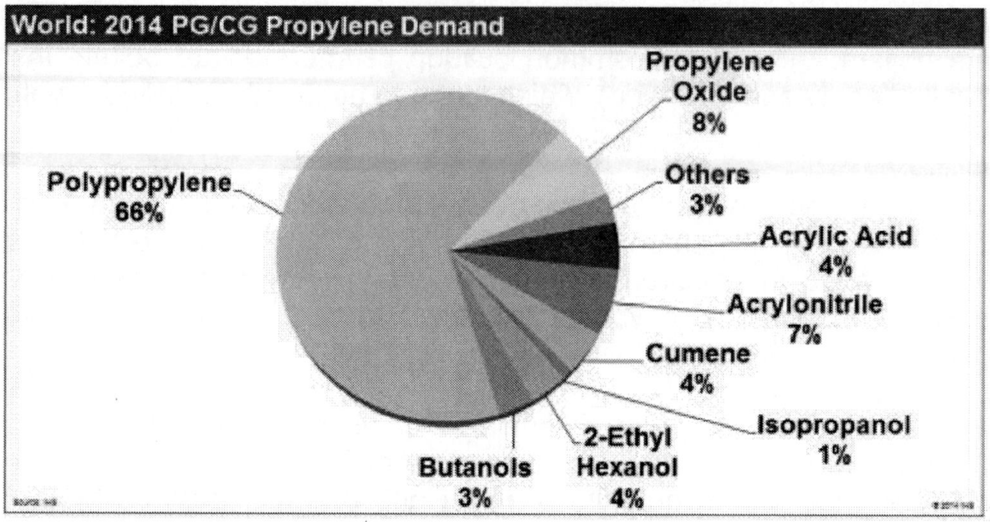

Polypropylene is a propylene-derived polymer that is widely used across the world in many different applications because of its favorable properties. In 2014 polypropylene manufacturing comprised 66% of propylene consumption. It is hydrophobic, chemically resistant, heat-resistant, fatigue resistant, and electrically resistant; it is lightweight, semi-rigid, and translucent; it has a high softening point, a high resistance to flexing stress, and a high impact strength. Even the most skeptical of sources agree that it is safe and recyclable. It has many applications, including water-resistant clothing, food and chemical packaging, tools, laboratory equipment, and a range of automobile parts.

Use of polypropylene is increasing, particularly in the automotive sector. Demand for more fuel-efficient vehicles has spurred the development of more lightweight polypropylene car parts. The transition of the developed world's car fleet to more fuel efficient vehicles and the expansion of the developing world's fleet will result in a high demand growth for polypropylene. The other large polypropylene demand growth sector is food packaging and containers, which is growing rapidly in China, India, Brazil, Southeast Asia, and Latin America. The combination of these two growth factors amounts to a forecasted 5.7% global CAGR from 2015 to 2023. In the long term, polypropylene demand growth will continue as consumer goods economies further develop in many parts of the world.

When first produced commercially in the 1950s with the discovery of Ziegler-Natta catalysts, polypropylene was produced in solvent. Technology has since grown to include liquid, gas, and supercritical polypropylene production in fluidized beds and stirred reactors – these advances have been facilitated by developments in catalyst activity and selectivity. Next-generation Ziegler-Natta catalysts and metallocene catalysts are allowing for lower production costs, higher purity, lower VOCs and most importantly, more control over structural and stereochemical properties. Global focus and investment in propylene and polypropylene technologies will continue to expand the potential applications and market share for polypropylene as new-generation catalysts continue to allow for more precise control over physical properties and overall improvement in key performance properties such as isotacticity and thermal resistance.

Increasing Worldwide Capacity

Worldwide capacity of polypropylene is set to increase dramatically with the global demand. Declining gasoline demand and the glut of shale gas will continue to spur growth in the chemical industry and put the U.S. in a position to be a net exporter of petrochemicals derivatives – a significant fraction of the polyolefins export will likely be in the form of solid resins.

The Middle East may also position itself to capitalize on the transition to petrochemical production. However, Middle Eastern oil exports may be well positioned to continue to supply growing Asian markets with gasoline and distillates, as China continues to grow steadily and India looks to grow even faster. Both regions will continue to import oil and cancel out decline in demand in the OECD countries. Unless China or India greatly expands its oil exploration, growing opportunity for Middle Eastern gasoline and distillate exports may soften the emphasis on petrochemical expansion.

FCC-Derived Polypropylene: Refinery Challenges

In the short term, supply for PP is tight and US demand is growing domestically. Polypropylene demand growth is expected to occur mostly in economies in which the consumer-goods sector is more rapidly expanding (China, India, Latin America); incremental growth in US-produced polypropylene will likely be exported in the long term, limiting the market to areas with marine port access.

The sale of chemical grade propylene to a polypropylene manufacturer may be limited by transportation and handling costs to make on-site or over-the-fence polypropylene production more feasible; concerns over propylene purity would likely prohibit the use of conventional pipeline infrastructure or require a dedicated pipeline for transport, limiting long-distance transportation of high grade propylene to rail or truck.

Onsite polypropylene production in a refinery would require only incremental capacity additions to the following common refinery utilities and offsites (U & O): polymer grade propylene and ethylene, hydrogen for chain termination, nitrogen, electricity, steam, cooling water, and water for post-resin wash.

Although refiners commonly handle solid catalyst, the handling of solid product at a much higher volume may present more extensive U & O expansion for solids handling equipment including pellet silos, extruders, pneumatic conveyors and truck or rail unloading, handling and storage of catalyst and polymer additives as well as nitrogen purge gas recovery and treatment to strip resins of VOC.

Lastly, no other refined products are sold in the plastics and resins market, presenting a potential market barrier for refiners and potentially necessitating a joint venture with a more seasoned resins producer or marketer.

In-depth analysis: Chinese Coal to Olefins May Be Done Growing

China looks to be one of the fastest growing markets for polypropylene and, until recently, one of the fastest growing producers via Coal to Olefins (CTO) technology. However, the Chinese long-term outlook on coal may lead to less of an impact than originally estimated.

China has recently set aggressive goals to reduce its dependency on coal in the name of cleaner air and carbon emissions reduction. China plans to close more than 1,000 coal mines this year, decreasing its coal capacity by 60 million tons, with a goal to cut production by 500 million tons of surplus production capacity within the next three to five years. More details of China's plan to cut coal capacity are outlined in the 13th Five-Year-Plan which was released on March 15th, 2016.

Although China intends to transition the bulk of its power generation capacity away from "dirty" coal, it will continue to use its vast coal infrastructure and workforce to make petrochemicals and "clean" coal energy sources via more efficient coal burning technologies, coal gasification, and synthetic fuel derivatives. In the petrochemicals sector, China has made significant investments in CTO technology over the past decade - Chinese CTO capacity was expected to total 12 million metric tons by 2016. The recently constructed facilities are expected to continue to become operational amid low crude prices. Although the economics of naphtha derived olefins are more favorable when crude prices are low, CTO production is less dependent on feedstock price than naphtha-derived olefins, and the Chinese government will prioritize the transition of coal workers to CTO and other coal-derivative facilities as they shrink the production and burning of "dirty" coal, particularly in the coal-rich western provinces.

The prospect of further growth in CTO and other coal gasification-related technology is up for debate. China will continue to weigh the cost of unemployed coal workers versus its commitment to cleaner environmental technology. From the standpoint of environmental politics, CTO and coal gasification are not favorable due to the intensity of energy and water consumption. The operation and possible expansion of CTO and polypropylene plants in the western provinces is economically favorable as labor costs rise in the eastern provinces and the country transitions away from coal, the economic staple of the western provinces. Additionally, China will upgrade infrastructure between the eastern and western provinces as part of its 5-year plan, enabling easier transport of olefins or polymers.

For now, there are mixed messages. Government reports from December indicate that future CTO projects will not be approved; however, the Chinese government has also announced continuing development of new syngas-to-olefin technologies. Whether or not CTO capacity will continue to grow, it appears that the PROC views CTO as an important and promising part of the future.

Conclusions

1) Petroleum refinery FCC units can be modified and switched from gasoline to high olefin mode to meet significant projected growth in propylene demand.
2) Operating in high olefin mode can increase FCC propylene yield by nearly an order of magnitude from a typical 3 to 5 wt% to over 25 wt%.
3) Significant modification and upgrading to the Reactor, Wet Gas Compressor, Gas Plant and other system components is required to produce and recover ultimate petroleum refinery FCC unit propylene production.
4) A Propylene-Propane Splitter and treatment system is needed to improve refinery grade propylene product purity to meet chemical and polymer grade specifications.
5) Production of propylene derivatives like polypropylene presents additional processing, handling and marketing challenges for petroleum refiners.

References

1. American Chemistry Council. (2014). What is Polypropylene Plastic? - Plastics Make It Possible. Retrieved March 17, 2016, from https://www.plasticsmakeitpossible.com/about-plastics/types-of-plastics/professor-plastic-one-of-the-plastics-superheroes-polypropylene/.

2. Business Standard. (2015, October 15). Rising demand for lightweight vehicles and packaging to boost polypropylene market. Retrieved March 17, 2016, from http://www.business-standard.com/content/b2b-plastics-polymers/rising-demand-for-lightweight-vehicles-and-packaging-to-boost-polypropylene-market-115100700616_1.html.

3. Carrington, D. (2016, January 19). China's coal-burning in significant decline, figures show. Retrieved March 17, 2016, from http://www.theguardian.com/environment/2016/jan/19/chinas-coal-burning-in-significant-decline-figures-show?CMP=share_btn_tw.

4. Ceresana. (2014, December). Market Study: Polypropylene (3rd edition). Retrieved March 17, 2016, from http://www.ceresana.com/en/market-studies/plastics/polypropylene/. Abstract.

5. Chicago Bridge and Iron Company. (2012). Polypropylene [Brochure]. Retrieved March 18, 2016, from http://www.cbi.com/images/uploads/tech_sheets/polypropylene-12.pdf.

6. Dean, C., "Naphtha Catalytic Cracking in FCC Units Past & Present ?" High Olefins Technology Services LLC. Presented at REFCOMM, Galveston, TX May 2015.

7. Dow Corporation. (n.d.). Dow Introduces Breakthrough Technology for Effectively Combining PE and PP. Retrieved March 17, 2016, from http://www.dow.com/en-US/news/press-releases/Dow Introduces Breakthrough Technology for Effectively Combining PE and PP#q=PP-Based Olefin Block Copolymers as Compatibilizers for PE and PP.

8. Duke University Nicholas School of the Environment. (2016, March 1). High Carbon Coal Products Could Derail China's Clean Energy Efforts. Retrieved March 17, 2016, from https://nicholas.duke.edu/about/news/chinaplanmayderailcleanenergyefforts.

9. Esposito, F. (2015, December 21). Profits, demand strong for PP market. Retrieved March 18, 2016, from http://www.plasticsnews.com/article/20151221/NEWS/151229979/profits-demand-strong-for-pp-market.

10. Jawad, Z., "Moving the FCC Out of Its Comfort Zone" Technip, Stone and Webster Process Technology. Presented at REFCOMM, Galveston, TX May 2015.

11. ICIS. (2007, November 6). Polypropylene (PP) Production and Manufacturing Process. Retrieved March 17, 2016, from http://www.icis.com/resources/news/2007/11/06/9076431/polypropylene-pp-production-and-manufacturing-process/.

12. IHS. (2016, February 29). US Bulk Chemical Exports to Increase in 2016 and Beyond as Shale Gas-Derived Production Expansions Come Online. Retrieved March 17, 2016, from http://press.ihs.com/press-release/chris-geisler/us-bulk-chemical-exports-increase-2016-and-beyond-shale-gas-derived-prod.

13. IHS. (1996). Supercritical Loop Reactor Slurry Process for Producing Polyethylene and Polypropylene [Abstract]. Process Economics Program Report, 19F. Retrieved March 14, 2016, from https://www.ihs.com/pdf/RP019F_toc_173554110917062932.pdf.

14. Letzsch, W. S. and Dean, C, "How to make anything with a catalytic cracker", Hydrocarbon Processing, July 2014.

15. Letzsch, W. S. and C. Dean, "Optimize olefins and aromatics production", Hydrocarbon Processing, November 2014.

16. Maddah, H. A. (2016). Polypropylene as a Promising Plastic: A Review. American Journal of Polymer Science, 6(1), 1-11. Retrieved March 14, 2016, from http://article.sapub.org/10.5923.j.ajps.20160601.01.html.

17. Marri, R. R., Dorsey, M., Portela, J. "Maximizing the Flexible Engine of the FCC for Meeting Propylene and Alky Feed Demands", Presented at REFCOMM, Galveston, TX May 2015.

18. Plotkin, J. S., & American Chemical Society. (2015, September 14). The Propylene Gap: How Can It Be Filled? Retrieved March 18, 2016, from http://www.acs.org/content/acs/en/pressroom/cutting-edge-chemistry/the-propylene-gap-how-can-it-be-filled.html.

19. Reuters. (2016, February 22). China to close more than 1,000 coal mines in 2016: Energy bureau. Retrieved March 17, 2016, from http://www.reuters.com/article/us-china-energy-coal-idUSKCN0VV0U5.

20. Richardson, J. (2016, March 7). China: Wrong Questions Will Give You The Wrong Answers. Retrieved March 17, 2016, from http://www.icis.com/blogs/asian-chemical-connections/2016/03/china-wrong-questions-will-give-wrong-answers/.

21. Schut, J. H. (2014, February 14). How New Catalysts Are Changing Polypropylene. Retrieved March 17, 2016, from http://plasticsengineeringblog.com/2014/02/14/how-new-catalysts-are-changing-pp/.

22. Sjolin, S. (2016, February 11). Move over, China - India will drive future oil-demand growth. Retrieved March 17, 2016, from http://www.marketwatch.com/story/move-over-china-india-will-drive-future-oil-demand-growth-2016-02-10.

23. Xinhua. (2016, March 15). China's new five-year plan boosts global sustainable development. Retrieved March 17, 2016, from http://english.gov.cn/news/top_news/2016/03/15/content_281475307954120.htm.

24. Yinqi., C. (n.d.). NPC deputy, a chemist, publishes breakthrough for oil industry. Retrieved March 17, 2016, from http://www.chinadaily.com.cn/china/2016-03/07/content_23764289.htm. China Daily.

444981 Stable Membrane Processes for Enhanced Olefin-Paraffin Separations in Refineries and Polyolefin Plant

Abstract:

Olefins are primary feedstocks for manufacture of petrochemicals. There is significant economic benefit in increasing (1) capacity of existing distillation operations while simultaneously reducing costs and (2) minimizing purge from polyolefin plants. Existing distillation for separation of olefins from paraffins (e.g., propylene from propane) is very difficult requiring large distillation columns and significant energy consumption. In principle membranes are small and should be easily retrofitable into distillation columns.

Membrane processes have been previously evaluated for separating olefins from paraffins. In many cases, silver salt facilitating agents have been incorporated into membranes to preferentially transport olefins. Historically with earlier membranes, while good separations have been initially demonstrated in the laboratory, membrane stability problems have prevented development of commercial systems.

Using customized amorphous fluoropolymer (CAF) membranes, Compact Membrane Systems has developed stable olefin-paraffin separation membranes with very good propylene flux and propylene/propane selectivity. The permeance and selectivity combination of the membranes are above the industry Robeson curve. In addition results were stable when exposed to high levels of conventional foulants (e.g., sulfur containing gases) in the laboratory. Using basic data from these laboratory results payback times for hybrid distillation-membrane processes were calculated to be less than 1-2 years in refinery operations. Similarly related to purge lines in polyolefin plants payback times are less than 12 months. Experimental data and modeling results will be discussed more extensively at the presentation.

444078 Streamlining the Simulation of a Propane Dehydrogenation Plant with Steady Catalyst Deactivation

Abstract:

With the current shale gas market, on-purpose dehydrogenation of propane has become a viable option for the production of propylene. The software ProMax® was used to simulate a commercial propane dehydrogenation plant that accounts for deactivation of the catalytic active sites along the reactor length. The plant includes four catalytic moving-bed dehydrogenation reactors with inter-stage heating, and a selective dehydrogenation reactor to saturate undesired species. Propylene is obtained by cryogenic separation and fractionation, and the unreacted propane is returned to the reactor section.

Mass and energy balance equations are applied for a reactor modeled with the reacting fluid flowing radially while the catalyst activity varies in the axial direction. A network was developed to include a complete set of reactions, including dehydrogenation/hydrogenation, and hydrogenolysis on the active metal sites, as well as cracking, cyclization, and oligomerization transformations on the active acid sites.

Since the catalyst is sent to a continuous catalyst regeneration section before being returned to the dehydrogenation reactors, the physical implication is that the deactivation profile for each type of site varies only along the length of each reactor and remains constant in the time domain. As a result, the deactivation of the catalytic sites was modeled by employing a recently developed empirical function dependent on the axial velocity of the catalyst and parameters that are intrinsic to the type of catalytic site.

The results of the simulation show that selectivity to propylene and per pass conversion obtained are comparable with commercial units, in the range of 85-92 wt.% and 29-35 % respectively. In addition, optimum conditions for the dehydrogenation reactors were determined to maximize the conversion of propane, and its selectivity to propylene. The impact on operating conditions and energy requirements of units downstream of the dehydrogenation reactors were also studied.

443215 WSA (Wet gas Sulfur Acid) Technology - A Competitive Solution for Sulfur Management

Abstract:

For more than 30 years Haldor Topsoe has designed plants for purification of sulfur containing waste gases derived from a wide variety of process industries. The WSA (Wet gas Sulfuric Acid) technology is efficient for recovering H_2S or other sulfur compounds in the form of commercial grade sulfuric acid. Moreover, the exothermic reaction energy generated when converting the sulfurous off-gas is efficiently recovered in the process as valuable superheated high pressure steam available for export.

The WSA technology is also characterized by high sulfur removal efficiency with more than 99% sulfur recovery, up to 99.99% when required, and is always in compliance with emission legislation. The process does not consume any chemicals nor will there be any solid or liquid effluents requiring additional treatment. This in combination with very low cooling water consumption makes WSA technology a very competitive solution for treatment of acid gases.

The WSA technology has proven feasible for treating sulfurous gases in many different applications, and today more than 140 units have been contracted worldwide handling a wide range of gas flows and with daily sulfuric acid production ranging from 4 to 1,140 MTPD.

The paper will explain the basic principles of the WSA technology and will include economical figures and the comparison with the typical alternatives for sulfur removal.

443717 Air-Quality Conscious Scheduling for Multiple Olefin Plant Start-Ups

Abstract:

The ground-level ozone is a pervasive air pollutant, which can be formed from flare emissions during chemical plant start-up operations. Especially in the concentrated industrial areas, the regional air quality (i.e., the ozone concentration) might be aggravated by simultaneously start-up emissions from multiple chemical plants. Thus, it is environmentally important for optimal scheduling of multi-plant start-ups under their manufacturing allowances to minimize the possible adverse air-quality impact. A systematic methodology on air-quality conscious scheduling for multi-plant start-ups has been developed. It demonstrates that multi-plant start-ups without any coordination could result in the significant air-quality impact. However, an optimal scheduling plan with several-hour tuning on the starting time of their start-up operations would significantly reduce about 80 % ozone increment. The study couples process dynamic simulation for industrial flare emissions with regional air-quality modeling together. It can provide valuable and quantitative supports for all relevant stakeholders, including environmental agencies, regional plants, and local communities.

444021 Integrated Waste Water Collection and Treatment System for Zero Waste Discharge Liquid Plant

Abstract:

Efficient design for collection of waste and effluent treatment is major challenge in EPC industry today. In this paper two cases of integration of waste water collection & treatment system are discussed. In first case, Power plant is considered and in other case LNG plant is taken. The sizing of waste water system and treatment of the effluents for zero waste discharge are compared for both of the cases. This paper throws light on significant savings by rightly sizing the collection and treatment system which majorly includes sumps, sump pumps, oil water separator, equalization tank, clarified water tank, and clarifier based on the intermittent and continuous waste from the plant. The flow sheets have been designed for power plant as well as LNG plant for integrated waste water collection & treatment system for zero waste discharge liquid plant. All the discharges like oil, waste water and chemicals are treated effectively and can be regenerated to use in the same plant. It is not only cost effective but also environment friendly.

BURNS McDONNELL

The Refinery "RTR" Rule – Major Refinery Impacts and Targeted Solutions

2016 AIChE Spring Meeting
Houston, TX
April 11, 2016

Blake Soyars, Section Manager, Environmental

Mark W. Lockhart, PE, Process Technology Manager Process & Industrial

Jim Corbit, Department Manager, Environmental

Greg Neve, Project Manager, Process & Industrial

Agenda

- What is the "RTR"
- "RTR" – Numerous Potential Impacts
- Long-Lead Compliance Projects Timing
- Roadmap to Success
- FCCU
  - Potential Effects and Solutions
- Delayed Coker:
  - Potential Effects and Solutions
- Flare Systems:
  - Potential Effects and Solutions
- Catalytic Reforming
  - Potential Effects and Solutions
- Benzene Fenceline Monitoring
- Conclusions

What is the "RTR"?

- "Risk & Technology Review" rulemaking only applies to petroleum refineries
- New US EPA Clean Air Act rulemaking finalized Dec. 1, 2015
- Five key rules updated
  - NSPS J & NSPS Ja focus on criteria pollutants VOC, SOx, NOx, CO & PM
  - MACT Y, MACT CC & MACT UUU focus on Hazardous Air Pollutants (HAP)
- Almost every US refinery will be affected
- Complex new requirements (following years of actual emissions reductions)
- Potential high initial compliance costs for many refiners:
  - EPA estimated **$2 million** per refinery average across 143 US refineries
  - American Petroleum Institute estimated **$8 million** per refinery average
  - Actual costs will depend on many site-specific factors
- Most costs will occur in 2016 - 2018 for overall Jan. 30, 2019 deadline

BURNS & McDONNELL 3

Refinery Risk & Technology Review (RTR): numerous potential impacts

- New/revised air emission standards drive compliance projects & add new requirements to certain economic projects.
- Each refinery needs detailed technical evaluations to identify, evaluate & address potential RTR impacts, including:
  - **FCCUs** – do regen pollution controls meet new requirements at all times?
  - **Reformers** – do formerly "exempted" regen vents <5 psig meet new requirements?
  - **Delayed Cokers** – are modifications needed for >2 psig decoking vents to atmosphere?
  - **Flares** – flare gas recovery and/or control/instrumentation upgrades needed?
  - **Storage Tanks** – new floating roof retrofits or roof fitting modifications required?
  - **Startup/Shutdown** – do emissions comply after RTR eliminates SSM exemptions?
  - **Overall** – does refinery have roadmap to evaluate all impacts and meet all deadlines?

BURNS & McDONNELL 4

RTR Timeline: long-lead compliance projects may be short on time

- For existing refinery sources, comply "as expeditiously as practical, but no later than…"
  - **Startup/Shutdown/Malfunction** – "exemptions" eliminated on Jan. 30, 2016 & replaced by "general duty" requirements, then new SSM limits apply by Aug. 1, 2017
  - **Storage tank standards** – implement by April 29, 2016 (can request more time)
  - **FCC emissions testing** – new tests/frequency with first new tests by August 1, 2017
  - **Benzene fenceline monitoring** – start by Jan. 30, 2018
  - **All other requirements** – comply by Jan. 30, 2019
- Will EPA revise the rule? minor "corrections" were proposed & industry will request more changes (outcomes uncertain)

Fluid Catalytic Cracking Units: potential project impacts for normal operation

- **RTR focus** for FCCUs is on the catalyst regeneration vent, which involves complex equipment that is extremely expensive to upgrade or modify
- **RTR impacts** are highly dependent on unit-specific configuration & operation: many units are ready to comply but detailed review needed for each unit
- **Normal Operating Requirements**:
  - Previous limits & requirements were selectively made more stringent
  - New testing by Aug, 1, 2017 & most other requirements apply Jan. 30, 2019
- **Startup, Shutdown & Hot Standby Requirements:**
  - Previous SSM "exemptions" replaced with new requirements matrix
  - Examples: new >20 ft/sec cyclone velocity for metal HAP & new 1% excess O2 limit for organic HAP
  - Some changes effective immediately: complete transition by Aug. 1, 2017

Fluid Catalytic Cracking Units: Solutions

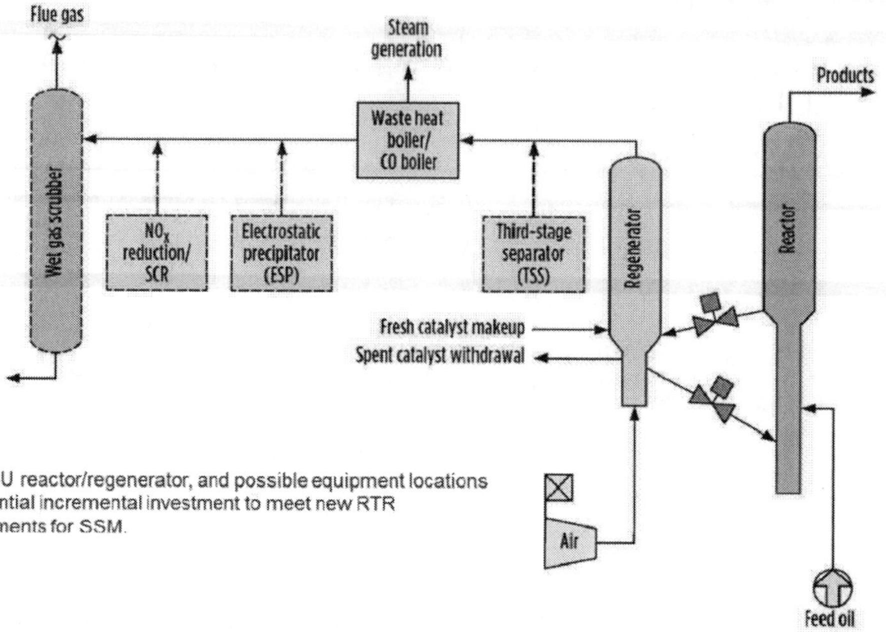

An FCCU reactor/regenerator, and possible equipment locations for potential incremental investment to meet new RTR requirements for SSM.

Fluid Catalytic Cracking Units – SSM

Third stage separators (TSS)	CO boiler (Partial burn)	Electrostatic precipitator (ESP)	Selective catalytic reduction (SCR)	Wet gas scrubber (WGS)	Comments:
				X	WGS assists with meeting PM, NO_x and SO_x requirements. CO met by complete combustion operation. Robust system.
	X				CO boiler assists with meeting CO. Evaluate incremental investment for TSS or ESP for meeting PM requirements on startup. SCR can assist with meeting NO_x.
		X			ESP assists in meeting PM but typically bypassed on startup. CO, NO_x and SO_x controlled by operating conditions in FCCU.
	X			X	CO boiler assists with meeting CO. WGS assists with meeting PM, NO_x and SO_x. Robust system.

Equipment exists

Equipment to consider as a minimum incremental investment

Delayed Coking Units: potential project impacts

- **Decoking vent pressure standards tightened for all delayed cokers:**
  - Old rule allowed drum venting as soon as pressure dropped to <5 psig outlet
  - New rule requires drum vapors to be processed or treated for organic HAP emission reduction down to <2 psig outlet
  - <2 psig limit applies to venting, draining, deheading & cutting
  - Additional technical requirement differences for "new" units versus "existing" units and for water overflow units versus other unit types
- **Emission reduction requirements unchanged:** all vapors that cannot meet <2 psig exemption must be recovered or treated to existing standards:
  - route to flare; or
  - reduce organic HAP by at least 98%; or
  - meet <20 ppmv Total Organic Carbon

Delayed Coking Units - Solutions

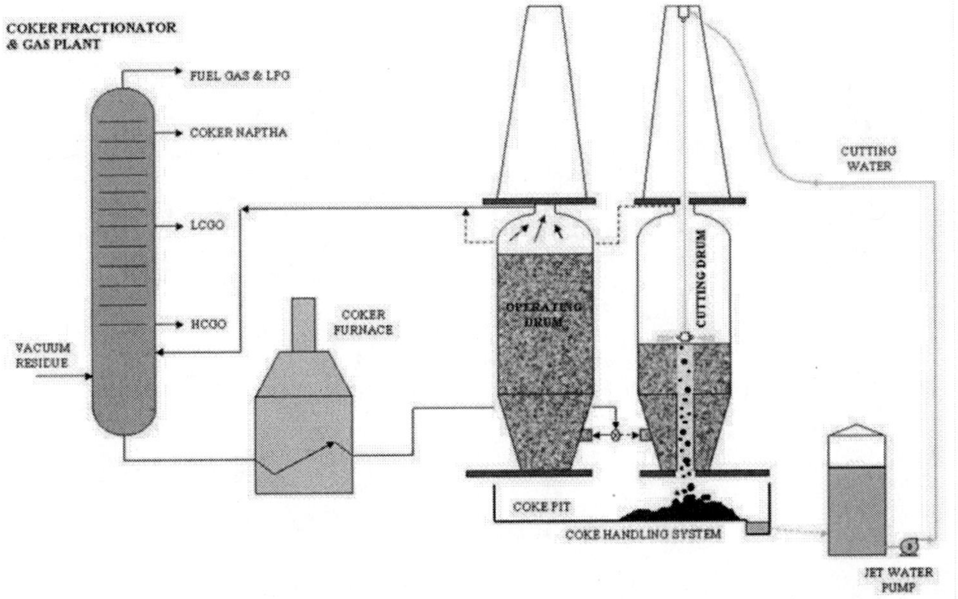

Source: Adams Project Managers, Inc.

Potential Mitigations

- **Achieving < 2 Psig Limit**
  - Use steam ejectors on blowdown drum
  - Use compressors on blowdown drum
  - Use compressors on coke drum overhead
  - Reduce pressure drop through blowdown system
  - Water filling in coke drums
  - Flare gas recovery compressor

BURNS & McDONNELL 11

Delayed Coking Units - Solutions

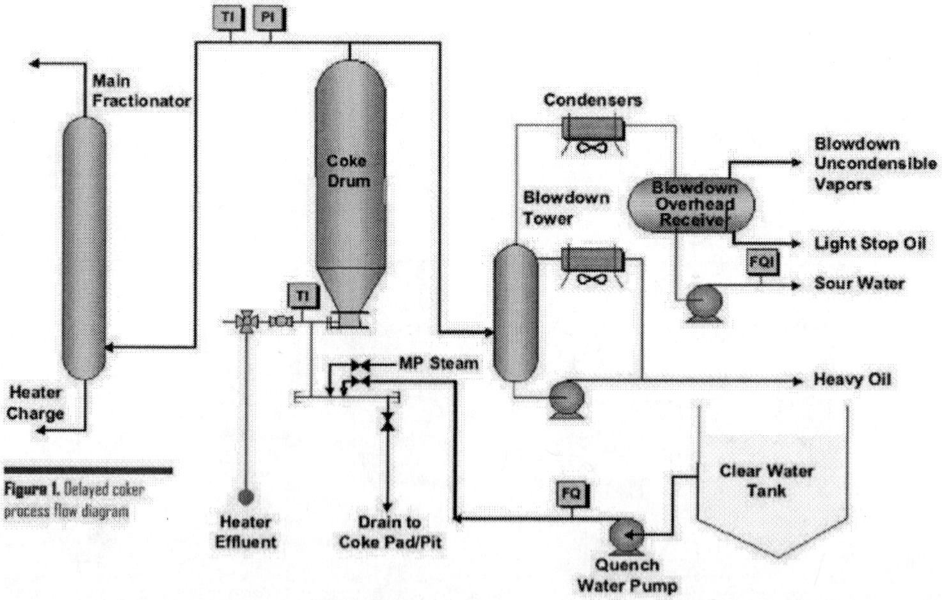

Figure 1. Delayed coker process flow diagram

Source: Lummus Technology, USA.

BURNS & McDONNELL 12

Design Considerations

- Can Coke Drums now achieve < 2 psig?
- Does achieving < 2 psig require more time which may affect capacity?
- What is the pressure drop through the blowdown system?
- Review increasing blowdown condensing capacity
- If add more motive force:
  - Ejectors – motive fluid; steam availability?
  - Compressors – Electrical capacity
- Main Fractionator – overhead cooling, compression and sour water handling.
- Consider a Flare Gas Recovery (FGR) system

Flare Systems: potential project impacts

- **RTR overhauls previous flare requirements**
- **New flare gas measurements required:**
  - Continuous volume flow measurement required for all "Combustion Zone" (CZ) contributing flows: vent gas, sweep, total steam, pre-mix, etc. (not pilot)
  - Quality of all flared gasses must be analyzed by instruments or sampling
- **Revised operating constraints for "normal" flare operation:**
  - New minimum Net Heating Value (NHV) ≥270 Btu/scf limit applies to all CZ flows
  - Tip velocity limits are similar to pre-RTR rules but apply to all CZ flows
  - NHV & velocity compliance calculations now required every 15-minute block
  - Startup/shutdown, excess fuel gas, PRV/PSV malfunctions, etc. regulated as "normal"
- **Emergency & high rates** >smokeless capacity subject to detailed new requirements
- **Bottom line**: RTR changes will drive detailed analysis & many flare projects

Flare System Solutions

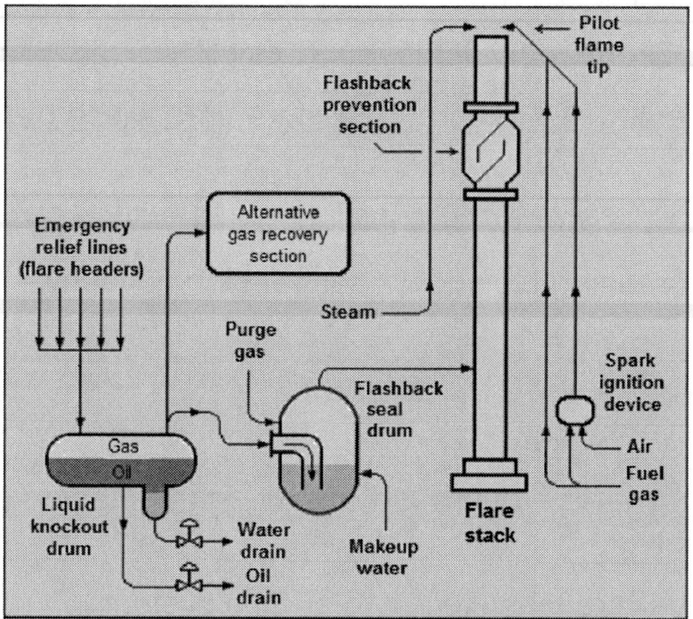

Diagram of an overall flare stack system in a petroleum refinery

Source: chemengineering.wikispaces.com.

Catalytic Reforming Units: potential project impacts

- **Scope of RTR changes for reformers:**
  - RTR only changes organic HAP requirements before catalyst burn-off
  - Inorganic HAP requirements for burn-off are unaffected by RTR
- **Old <5 psig vent "exemption" narrowed with more units regulated:**
  - Exemption now "clarified" to only apply to passive vessel depressurization
  - Any active purge vent (even if <5 psig) with sufficient pressure to route into emission treatment must now be treated to remove organic HAP
- **Organic HAP standards unchanged:** previously "exempted" <5 psig vents can no longer be vented unless routed to control: flare or other device to meet 98% HAP reduction or <20 ppmv TOC

Catalytic Reforming

Continuous Catalytic Regeneration (CCR) Units

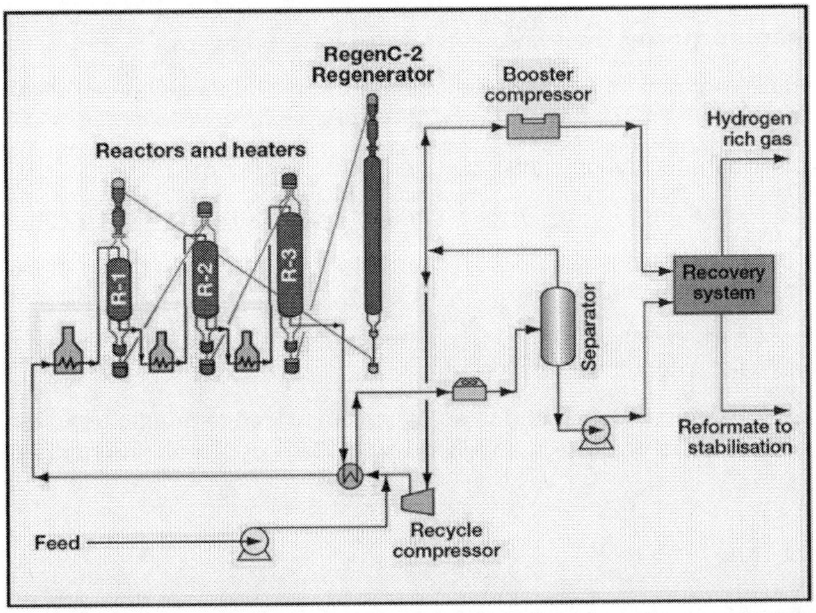

Source: Axens IFP Group Technologies.

Catalytic Reforming

Semi-Regenerative Units

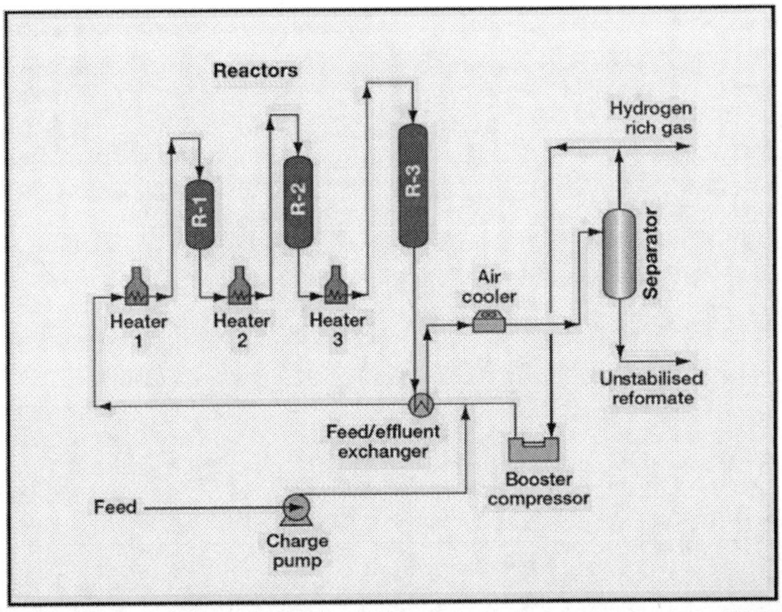

Source: Axens IFP Group Technologies.

Benzene Fenceline Monitoring Requirements

New program required to measure ambient benzene around refinery:

- Install 12 to 24 permanent fenceline/perimeter monitors (depending on refinery footprint) plus any background monitors to identify off-site benzene contributions
- Passive absorbent technology detects very low concentrations 24/7
- Meteorological station also required but can use existing/nearby met station
- Benzene data must be disclosed to public with associated new community perception & activist empowerment challenges
- Conduct root cause analysis if annual average measured benzene is >9 μg/m3 after discounting off-site contributions
- Multiple high measurements with related causes may lead to project requirements or increased compliance costs e.g., enhanced piping leak detection and repair program

BURNS & McDONNELL 19

Benzene Fenceline Monitoring Example

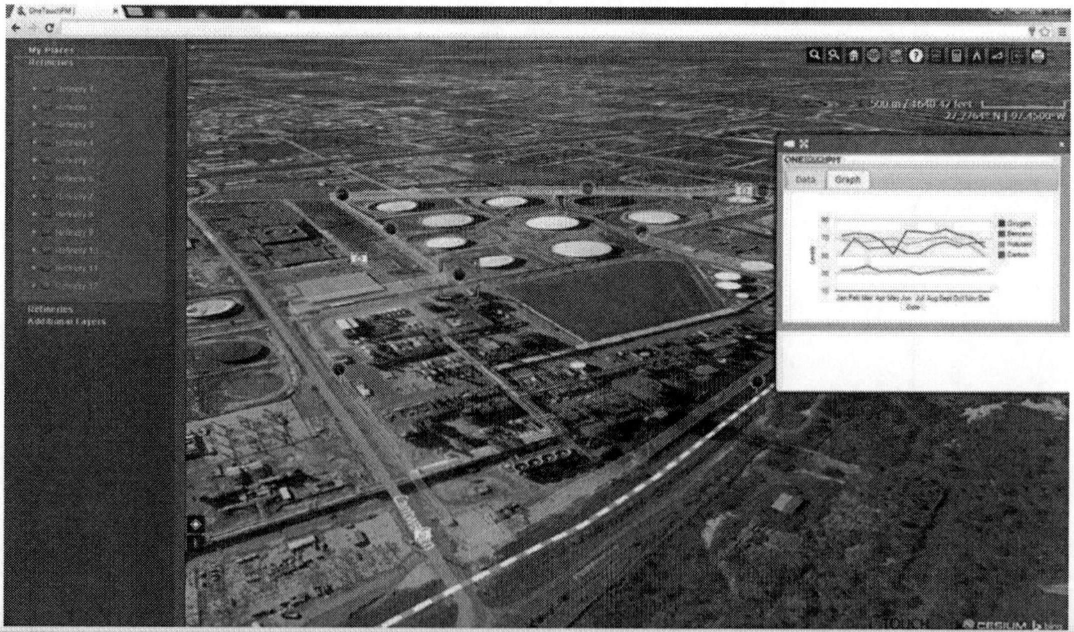

BURNS & McDONNELL 20

Other Impacts

- **Sulfur Plant**: new S/U & S/D requirements for purging gas
- **Storage Tanks**: some existing tanks may need new floating roofs, some existing floating roof fittings may need upgrades with April 29, 2016 deadline
- **Relief Valves:** new equipment requirements so that every release has: immediate detection, time/duration measurement, and operator notification
- **Maintenance Vapor Clearing:** new requirements apply even to small equipment clearing: affects procedures & compliance documentation
- **Onstream Analyzer Vents**: are now regulated process vents requiring vent-specific evaluations (many should meet flow/concentration limits)
- **Marine Vessel Loading**: certain loading was exempt from submerged-fill requirements but those exemptions were eliminated

Floating Roof Tanks, and Other Impacts:

External Floating Roof Tank (Pontoon Type)

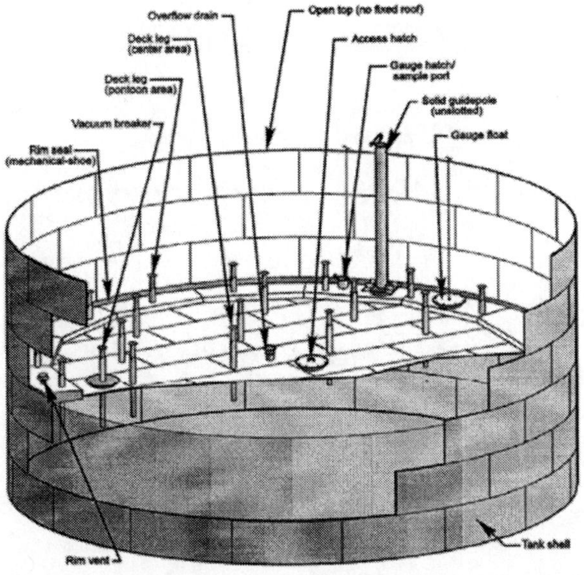

Source: US EPA, AP 42, November 2006.

Conclusions/Recommendations

- **Complexity:** is not reflected in slides, see paper for more detail
- **Priority**: plan for significant RTR compliance implementation costs with typical costs estimated in $2-8 million range per refinery
- **Scope**: the RTR can impact key refinery units and drive complex projects that may need coordination with turnarounds or other projects
- **Urgency**: the Jan 2019 overall compliance deadline combines with earlier deadlines for certain individual requirements to create schedule difficulty
- **Roadmap** is recommended to reflect priority, scope & urgency to chart course for successful compliance implementation with managed impacts
- **Team**: a purpose-built RTR implementation team is recommended to engage many necessary disciplines & stakeholders
- **Success** depends on site-specific RTR roadmap, implemented by an integrated RTR team, aligned with commercial priorities, and driven by compliance commitment

BURNS & McDONNELL 23

Contact Information

Blake Soyars, PE - Author/Presenter

Austin | 512-872-7150 | brsoyars@burnsmcd.com

Mark Lockhart, PE - Author/Presenter

Houston | 832-214-1975 | mlockhart@burnsmcd.com

Jim Corbit - Author

Houston | 832-214-2818 | jrcorbit@burnsmcd.com

Greg Neve - Author

Houston | 832-389-5746 | gneve@burnsmcd.com

BURNS & McDONNELL

CREATE AMAZING.

444253 Modeling a Low-Severity Ethylene Oligomerization Process

Abstract:

In order to take advantage of low feedstock costs, the Synfuels International / Byogy Renewables Biojet process uses crude ethylene as an intermediate to produce heavier olefins. This oligomerization process is challenging to operate and simulate. The highly exothermic nature of the reaction necessitates a heat-exchanger reactor design and multiple sections. In this reactor, a highly active catalyst is chosen to convert the ethylene into intermediate length olefins. While the distribution is not as important as the conversion, longer average chain length leads to more deactivation due to waxy buildup. Because of this, the shortest oligomer, butylene is targeted.

The modeling of this process creates the potential for a large number of species, both in terms of hydrocarbon length and type. In addition, the number of potential reactions (oligomerization, cracking, hydrogenation, aromatization, etc.) is substantial. It is desired to rapidly produce a model of the process to enable process design and optimization of larger scale reactors from existing lab and pilot scale data. The scope of these reactions is included in the kinetic technology of Bryan Research & Engineering, Inc.'s AutoKinetic Reactors in their ProMax simulator. This allows the user to easily test a large number of combinations of reactor designs and conditions. In addition, with experimental data, the reactor model can be tuned to more closely replicate the process catalyst and conditions. In this way, the process has been studied to give a reactor design with a minimum capital cost for a given conversion or yield, and to give maximum yield for a given reactor layout. This study has also identified and mitigated temperature and composition sensitive regions of the reactor.

443917

APPLICATION OF DIVIDING WALL / THERMALLY COUPLED DISTILLATION COLUMNS

11th April 2016

Sekhar Babu Mamilla, Antonio Matarredona, Christoph Roosen, Radu Ignat Michael Schleger
Chemicals Technology, Feedstocks and Olefins Europe, SABIC, the Netherlands

CHEMISTRY THAT MATTERS™

CONTENTS

- Conventional Technology
- Technical Background of Thermal Integration in existing distillation trains
- Requirements and Limitation
  - Possible solutions to eliminate the limitations
- Results summary
- Conclusions and Recommendations

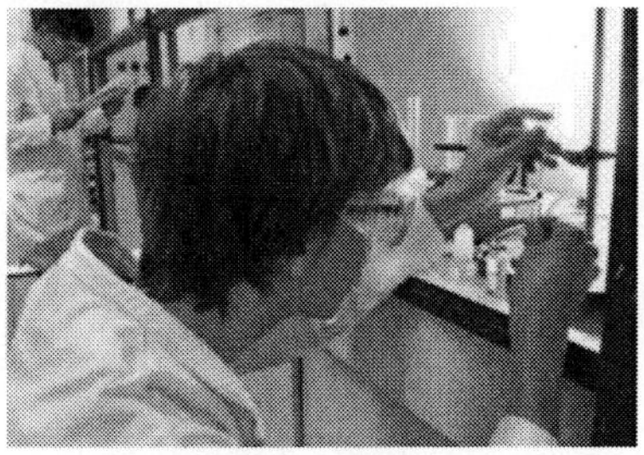

CURRENT STEAM CRACKER TECHNOLOGY

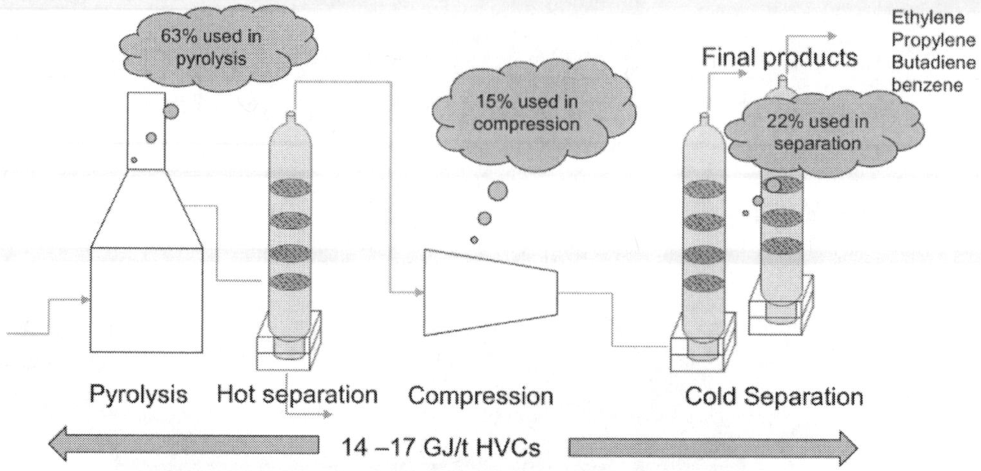

- Total World wide CO_2 Emissions are 180 – 200 million tons
- Theoretical thermodynamic energy requirement is 5 GJ/t HVCs
- Remaining amount is lost to the environment and products
- Distillation is thermodynamically one of the most inefficient process due to exergy destruction

Olefins from conventional and heavy feedstocks Tao Ren et.al. Energy Volume 31, Issue 4, March 2006

CONVENTIONAL PROCESS

THIS SEQUENCE COMES WITH INHERENT THERMODYNAMIC INEFFICIENCY

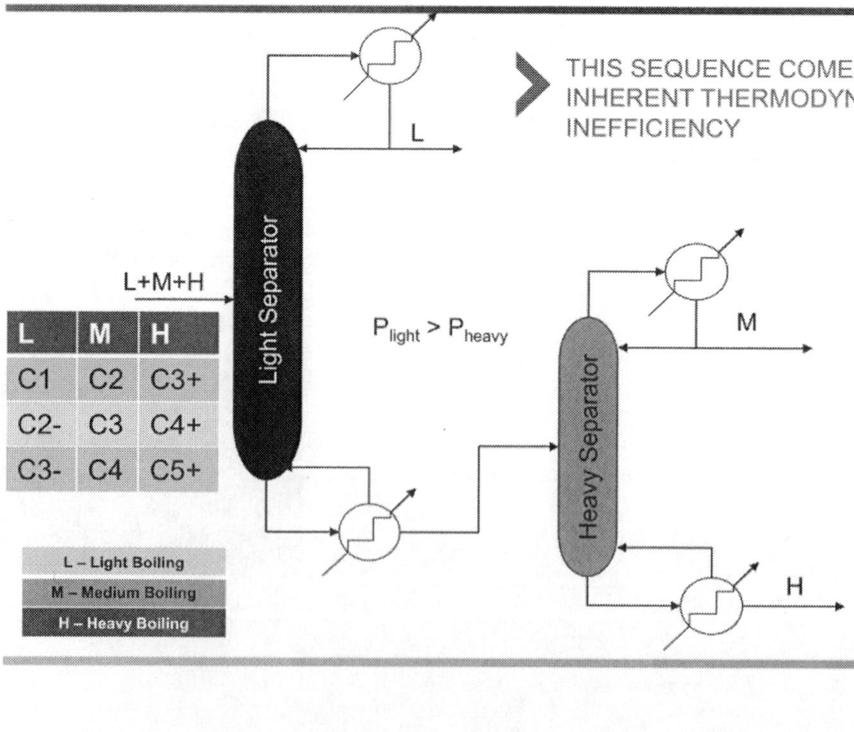

L	M	H
C1	C2	C3+
C2-	C3	C4+
C3-	C4	C5+

REMIXING EFFECT OF MEDIUM BOILING FRACTION

Concentration profiles of Medium Boiling fraction

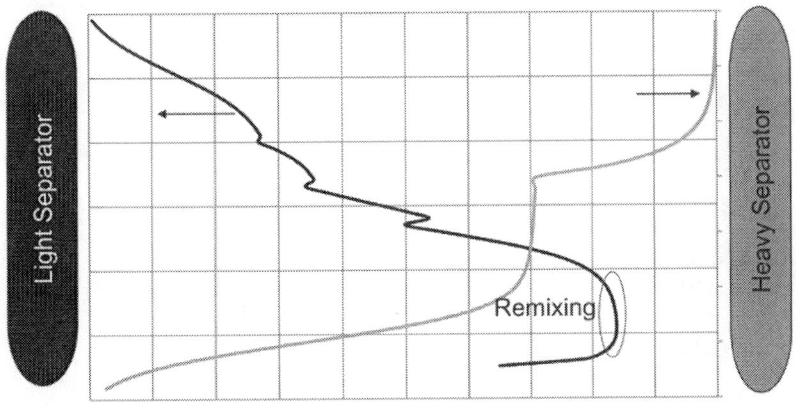

REMIXING DECREASES THE THERMODYNAMIC EFFICIENCY

No. 4

PRINCIPLE OF THERMAL INTEGRATION

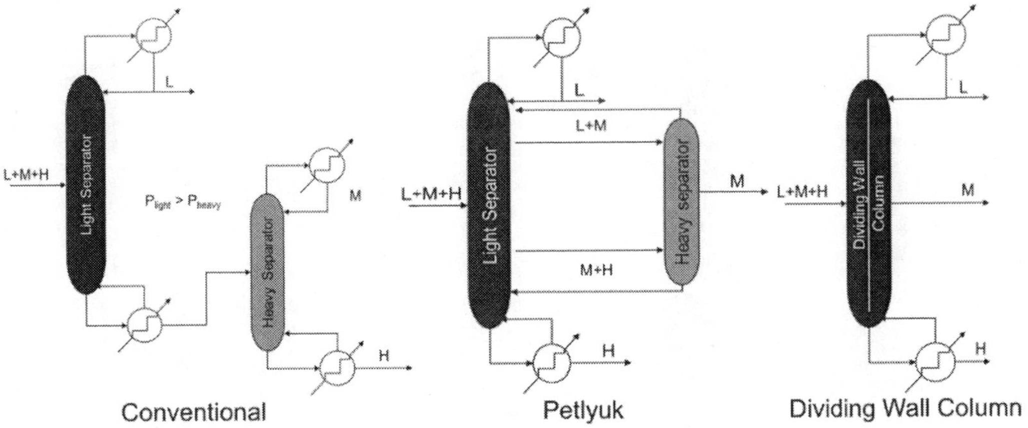

EVOLUTION OF THERMAN INTEGRATION

No. 5

VARIATIONS IN THERMAL INTEGRATION OF EXISTING COLUMNS

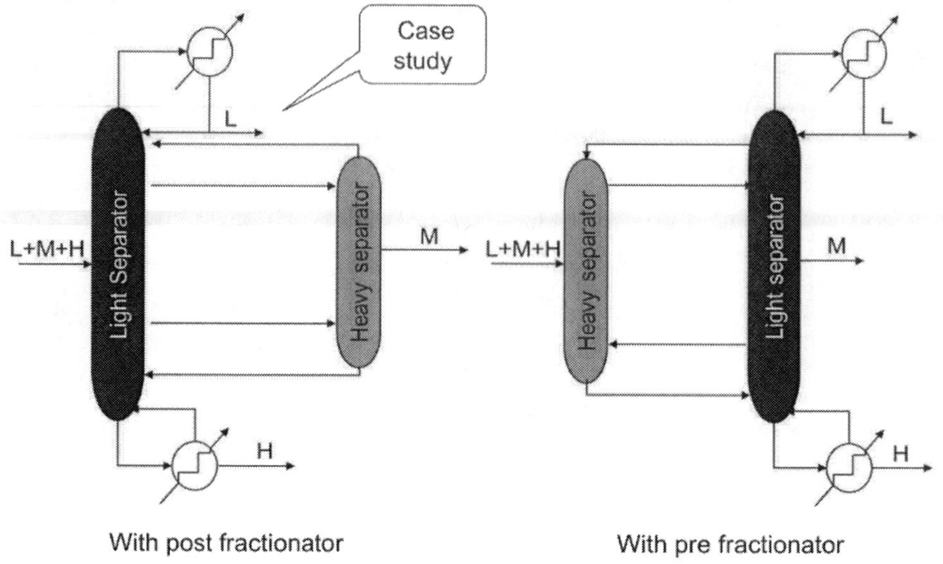

No. 6

REMIXING REMOVAL WITH POST FRACTIONATOR

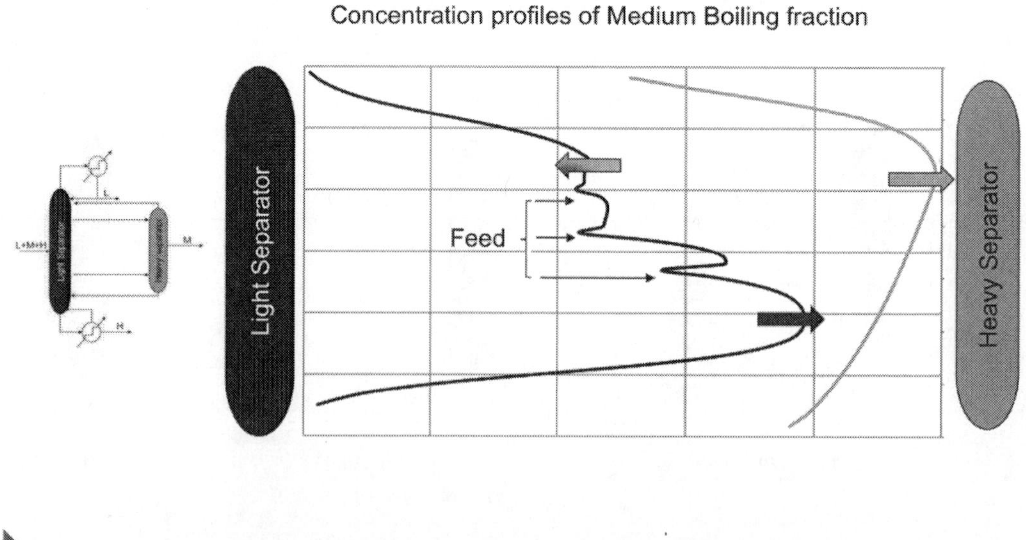

- NO REMIXING IMPROVES THE THERMODYNAMIC EFFICIENCY

No. 7

CONDITIONS AND LIMITATIONS OF THERMAL INTEGRATION

- Thermal Integration is practical if both the columns operate at similar pressure levels
- Otherwise, need to bring them to common pressure
- This will either increase the bottom temperature or decrease the top temperature
- If common pressure is not possible:
  - Small compressor
  - Partial thermal integration; like side draw

IDENTIFYING THE RIGHT COLUMNS IS THE KEY IN THERMAL INTEGRATION

No. 8

THERMAL INTEGRATION USING A COMPRESSOR(PTCC-1)

COMPRESSOR IS USED TO RECYCLE THE LIGHTS BACK TO 1ST COLUMN

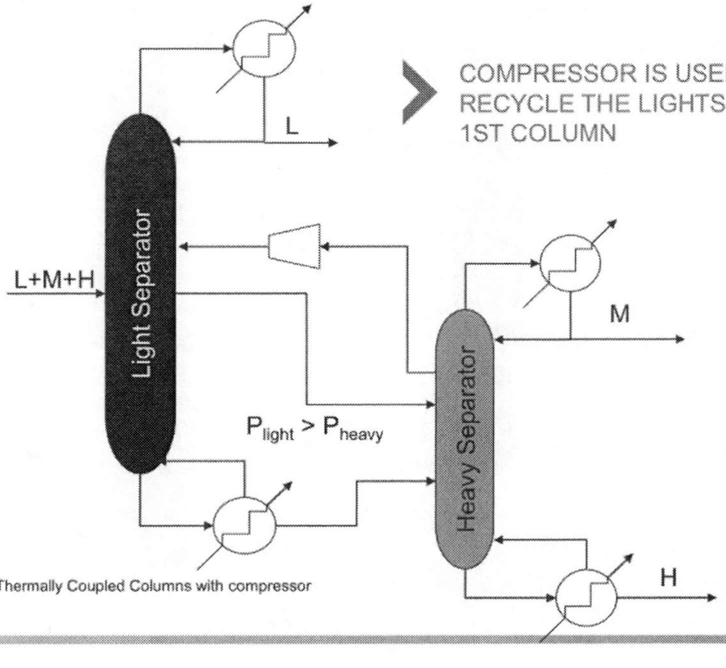

PTCC-1 : Partial Thermally Coupled Columns with compressor

No. 9

POSSIBILITY OF M.B.P SIDE DRAW TO INCREASE EFFICIENCY

Concentration profiles of Low and Medium Boiling fractions

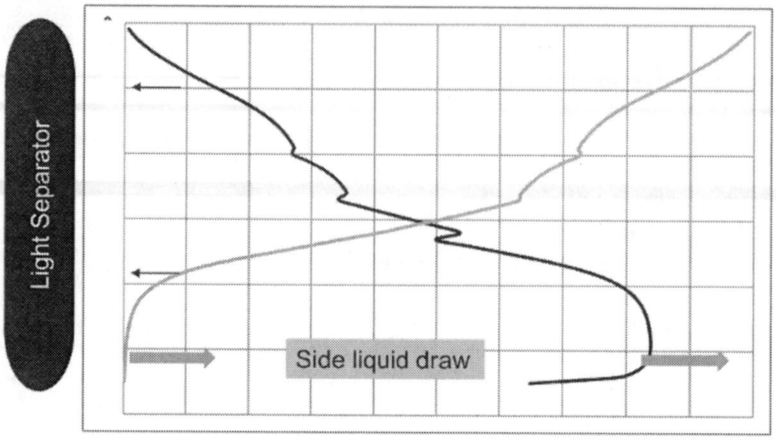

- MEDIUM BOILING SIDE DRAW WILL INCREASE THE EFFICIENCY

No. 10

THERMAL INTEGRATION WITH A SIDE DRAW (PTCC-2)

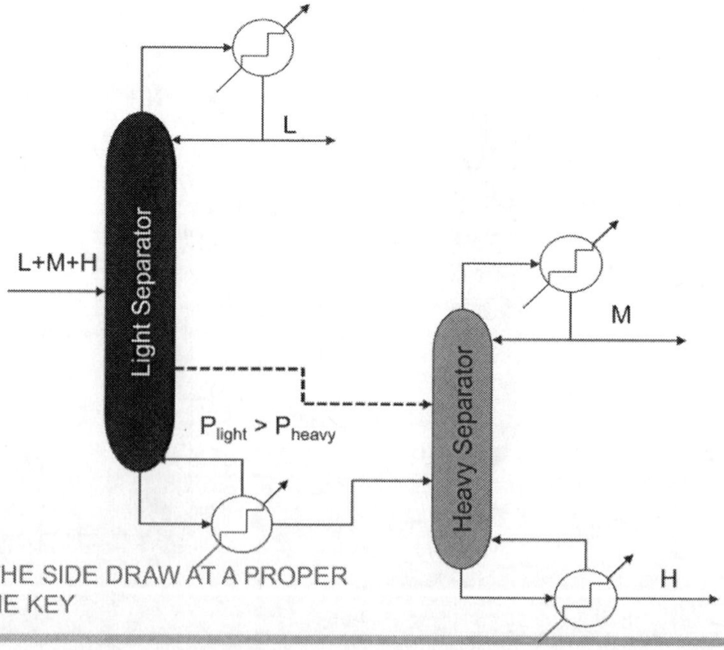

- PLACING THE SIDE DRAW AT A PROPER TRAY IS THE KEY

No. 11

REMIXING REDUCTION WITH SIDE DRAW

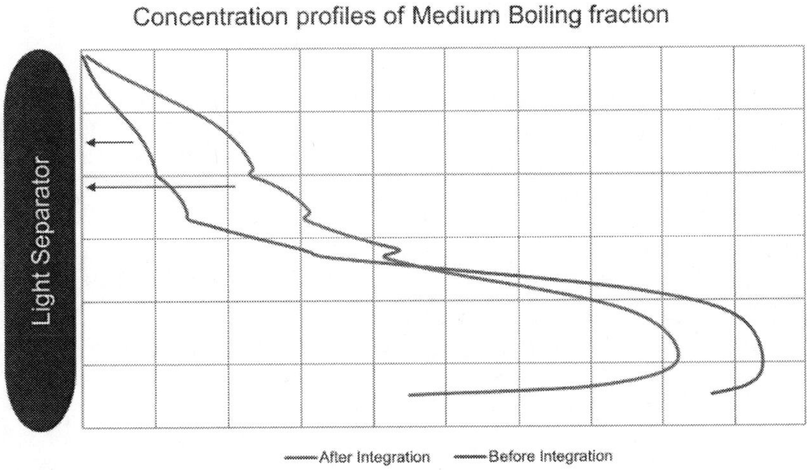

PROPERLY LOCATED SIDE DRAW REDUCES THE REMIXING

No. 12

SUMMARY OF THE RESULTS

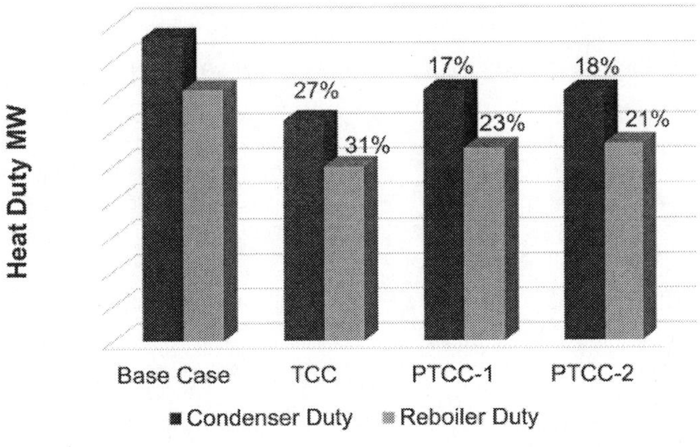

TCC : Thermally Coupled Columns
PTCC-1 : Partial Thermally Coupled Columns with compressor
PTCC-1 : Partial Thermally Coupled Columns without compressor

DEGREE OF THERMAL INTEGRATION IMPACTS THE SAVINGS
UTILITY SAVINGS HAS TO PAY FOR THE COMPRESSOR

No. 13

SWOT ANALYSIS

STRENGHTS	WEAKNESSES
• Proven Industrial technology • Significant Energy Savings • Minimum investment • Reverting back to standard operation is easy with a proper control	• Common pressure requirement • Tray and other modifications needed
OPPORTUNITIES	**THREATS**
• Can increase capacity or remove current bottleneck	• Compatibility of new configuration to large composition variations in the Feed • Increased bottom temperature can increase the fouling reaction rates • Operational Integrity with increased ΔP of one column

No. 14

CONCLUSIONS AND RECOMMENDATIONS

Current status:

- Hydrodynamic rating of the columns and other modifications needed to determine the investment

Conclusions :

- Industrial proven process intensification technology
- Low risk associated with the technology
- Can be applied to remove bottlenecks and/or increase capacity

Recommendations:

- Identify an appropriate section in the plant for thermal integration to maximize savings
- Depending on the savings and the pressure difference; compressor can be used
- Conventional side draw can be applied to improve the thermal efficiency of the process
- Smart control structure can be applied to improve the integrity of such solutions
- Online analyzers can be used to build the smart controller

No. 15

سابك
sabic

THANK YOU

Sekhar Mamilla
Scientist
Chemcal Tecnnology Europe
sekhar.mamilla@sabic.com
www.sabic.com

No. 16

16AIChE
Spring Meeting, Houston, TX

The Principle of Intrinsic Reality
Solving the World's Woes through Chemical Engineering and Economics

Jed Moyano Bellen, SCPM, PMP
Hyundai Engineering & Construction Co. Ltd.
9 Temasek Blvd., #27-02 Suntec Tower 2, Singapore 038989
jed.bellen@hdec.co.kr

Prepared for Presentation at
American Institute of Chemical Engineers
2016 Spring Meeting
11th Global Congress on Process Safety
Houston, Texas
April 10-14, 2016

UNPUBLISHED

The Principle of Intrinsic Reality Solving the World's Woes through Chemical Engineering and Economics

Jed Moyano Bellen, SCPM, PMP
Hyundai Engineering & Construction Co. Ltd.
9 Temasek Blvd., #27-02 Suntec Tower 2, Singapore 038989
jed.bellen@hdec.co.kr

Keywords: Principle of Intrinsic Reality, Chemical Engineering, Economics, Sustainability

Abstract

Human beings in profound rationality always seek for explanations of the phenomena around them. To understand these phenomena, human beings have developed systematized body of knowledge as a practical way of dealing with life's challenges. Among the systematized body of knowledge developed are chemical engineering and economics.

An avant-garde way of looking at each of these disciplines leads to the Principle of Intrinsic Reality. This principle relates the fundamental laws of the two disciplines to glean insights on the constraints imposed by conservation laws on economic systems at both the macro and micro levels.

It can be argued that if mass and energy in the universe are constant, then logically diminishing returns and marginal utility are bound to be observed. This logic is quite evident in a microorganism culture which follows the Monod growth model.

It can also be thought that conservation laws imply scarcity of resources. As such, it enjoins everyone to find optimal and efficient use of resources in the context of trade-offs, production decisions and sustainability. It even calls for the development of new technologies to produce more outputs from less inputs (i.e. shifting of production possibility frontier to the right). In this way, the Principle of Intrinsic Reality leads to the imperative for sustainable optimization.

Furthermore, it is deemed that the conservation laws can explain the phenomenon called "polarity of wealth." As mass and energy when harnessed create wealth, then the constancy of mass and energy implies boundary for wealth creation. In effect, wealth can neither be created nor destroyed but can only shift from one entity to another thereby creating polarity in wealth distribution. This assertion can be the key towards more equitable restructuring of economic systems to close income gap and thereby result to stable and resilient populace.

In conclusion, this paper envisions for a paradigm shift from traditional money-centric view of chemical engineering and economics to a synergistic convergence of disciplines which can powerfully tackle the world's woes and thereby effect sustainability.

1. The World in which We Live Today

The past few years saw unprecedented global challenges which have put into question existing economic models and scientific paradigms. Among these challenges are the dystopian global economy, climate change, environmental degradation and terrorism. The magnitude of these challenges has grown to a profound level that newer ways of thinking and perspectives become necessary. With that end in mind, this paper attempts to look at the fundamental laws of chemical engineering and economics in new ways to recognize our intrinsic reality from which these pressing challenges can be tackled at large.

2. The Principle of Intrinsic Reality

It is an observable fact that advancement in science increasingly blurs boundaries of various disciplines in tackling proactively the pressing global issues or problems. Within this purview, two seemingly diverging disciplines namely chemical engineering and economics can be taken together to glean insights on our intrinsic reality and what it means for our economic and ecological systems on a larger scale.

In the author's contention, intrinsic reality is grounded on the fundamental laws of chemical engineering and economics. These laws are the Conservation Laws (Chemical Engineering) and Law of Diminishing Returns (Economics). It can be argued that the Law of Diminishing Returns is observable because of the constraints imposed by the Conservation Laws. As such, this leads to the Principle of Intrinsic Reality which can be stated thus;

> *In any system, both at the macro- and micro-levels, the limiting factor for growth is the constancy of potential inputs.*

Potential inputs mean not the actual inputs to processes but the constant material or energy available which can be harnessed to create wealth. The reasoning or proof of this principle is presented in subsequent discussions.

2.1 Proof Based on the Monod Growth Model

The best illustration to understand the Principle of Intrinsic Reality is the Monod Growth Model. Monod in his 1949 seminal paper presented an empirical relation between exponential growth rate and concentration of essential nutrient. The data was fitted into hyperbolic equation:

$$R = R_K\left(\frac{C}{C_1 + C}\right)$$

In this equation, R is the growth rate during the exponential phase; R_K is the rate limit for increasing concentrations of C; C is the concentration of the nutrient; and C_1 is the concentration of nutrient at which the rate is half the maximum.

Figure 1: Graphical Representation of Monod Growth Model

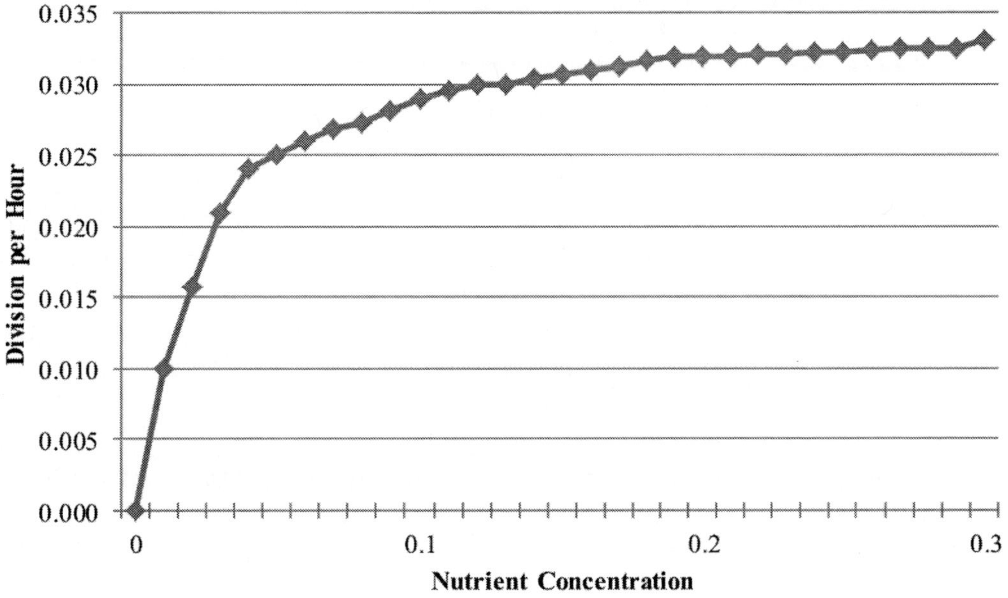

Monod (1949) further pointed out that the limiting factors commonly found are exhaustion of nutrients, accumulation of toxic metabolic products and changes in ion equilibrium.

The foregoing discussion has been observed at the micro-level with the bacterial culture. At the macro-level (i.e. the world at large), analogous phenomena are also observed to occur because of the same constraint imposed by the constant quantity of resources (i.e. material and energy). One example of these phenomena is that human beings cannot grow indefinitely. If they do, then how can the world sustain that growth with limited available resources (e.g. foods) to support it?

Another analogous phenomenon is in population growth. Thomas Malthus (1789) predicted that when population growth exceeds agricultural production, the population level is forced to return to subsistence-level conditions. This phenomenon as Malthus (1789) believed is due to the restriction of available resources supporting population growth, which others have claimed to be exploding as the earth becomes over populated.

In a surprising manner, a similar phenomenon can be observed in production processes in which the increase in a single factor of production assuming all other factors constant, results not to more marginal output but to a reduced one. This was attributed by classical economists, Thomas Malthus and David Ricardo, to the decreasing quality of inputs while neoclassical economists attributed this to the disruption in the entire production process as additional units of input are added to a fixed amount of capital. But in view of the Principle of Intrinsic Reality, the phenomenon is viewed intrinsic to the production system in which case infinite amounts of mass or energy cannot be processed with perfect efficiency. Hence, the law of conservation of mass and energy comes into the fore.

It can be deduced from the discussion that there is an underpinning intrinsic reality to the more obvious phenomena in the world from the micro-scale (bacterial growth) to the macro-scale (human growth and production processes).

2.2 *Proof Based on Diminishing Marginal Utility*

Another way to prove the Principle of Intrinsic Reality is through the diminishing marginal utility observable among humans and perhaps other organisms as well. Marginal utility, as defined in economics, is the gain from an increase or the loss from a decrease in the consumption of goods or services. The diminishing marginal utility is an observable phenomenon in which the utility or satisfaction an individual gains diminishes as he or she consumes more of the same unit of goods or services. Graphically, this can be represented as in Figure 2.

Figure 2: Graphical Representation of Diminishing Marginal Utility

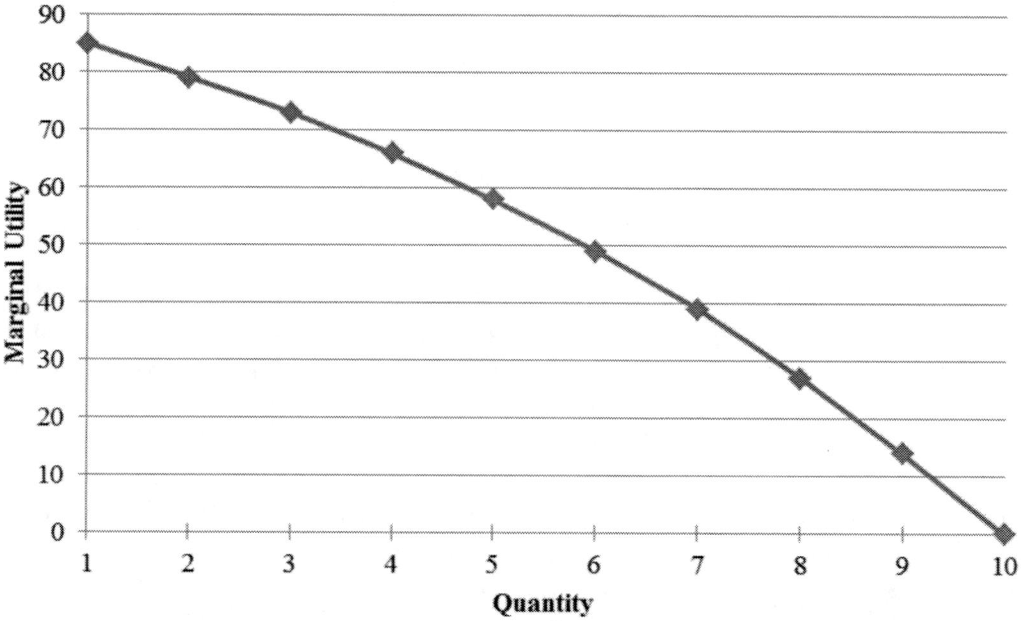

It can be deduced from this phenomenon and in support of the discussion in section 2.1 that human beings cannot grow indefinitely due to the fact that in our intrinsic reality, there is always a limit so as the constancy of available resources (i.e. mass and energy) can support our existence.

This means that in view of our intrinsic reality, we cannot have infinite greed otherwise an upsurge of unintended consequences are bound to happen (i.e. environmental degradation, terrorism and eventually demise of humanity).

Therefore, recognizing our intrinsic reality is the first step towards solving the world's woes and thereby resulting to sustainability and security.

2.3 *Expanding the Production Possibility Curve through R&D*

The caveat here is that the constant mass or energy (i.e. the limits) leads to trade-offs. Trade-offs is encountered when personal or business decisions are made. In making these decisions, one has to ensure that he or she becomes better off with his or her chosen option. We also realize in our life that we cannot get all that we want and we have to make trade-offs so long as it places us in a better or preferred position. This is more of a philosophical discourse, but it goes to show that the concepts of chemical engineering can be as useful in life and in business as it is in chemical processing.

To tackle these trade-offs, the Production Possibility Frontier in economics can be used. In view of the Principle of Intrinsic Reality where growth is limited by the fixed amount of material and energy, the Production Possibility Frontier needs to shift to the right as shown in Figure 3.

Figure 3. Graphical Representation of Expanding the Production Possibility Curve

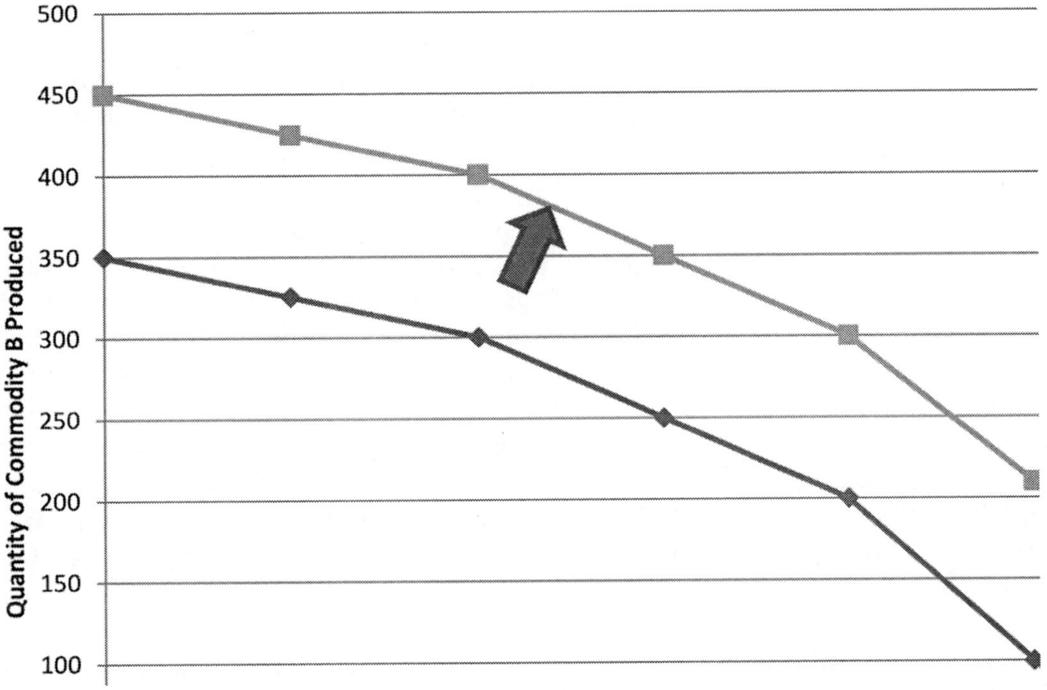

This shift has an implication on the practice of chemical engineering as Research and Development (R&D) is needed to achieve the said shift. R&D enables the economy to produce more of the two commodities (i.e. Commodity A and Commodity B) resulting in economic growth. The key challenge here to Chemical Engineering, particularly on Chemical Engineering R&D, is to look at the bigger picture where economic growth emanating from R&D output is pivoted on sustainability; considering its impact to the environment and the society at large.

3. Paradigm Shift in the Practice of Chemical Engineering

"Money" or "Profit Maximization" has been at the center of the practice of chemical engineering (Smith 2005). Various mathematical optimization methods are used to design systems, processes or components of processes with the end in mind of optimizing profits. This mental model however, will no longer work in our situation today as we face economic, ecological, social and spiritual crisis (Scharmer 2010).

To address this, Scharmer (2010) pointed out seven acupuncture points to shifting capitalism into a regenerative ecosystem economy. Within this purview, the practice of chemical engineering has a greater role to assume. For example, Scharmer (2010) suggested that the design of all production and consumption cycle be completely earth-to-earth, i.e. without necessitating for landfill and that it has to co-evolve with the natural ecosystem. In this regards, the way we do optimization should be revamped into sustainable optimization. By sustainable optimization, it means optimization of the entire system and adding in environmental as well as societal objectives to the equation.

In this definition, two things need to be considered. The first one is systems thinking which emphasizes the understanding of the entire system so the interaction among components will be known and unintended consequences mitigated (Senge 2006). This is a holistic approach to problem solving and finds enormous use in process design.

It should be noted however that the construction of processing plant, whether it is a greenfield or a brownfield project, is rooted on projected market demand. In some instances, economic volatility necessitates companies to experiment with multiple pathways to survive. In this circumstance and within the purview of systems thinking, design thinking comes into the fore. Design thinking is a collective principle emphasizing empathy for users, doing prototyping and tolerance for failure (Kolko 2015). This principle finds enormous use in chemical engineering designs such as in product development, process development, systems improvement and the like. This is also useful for simplifying or humanizing chemical engineering designs with the end in mind of contributing to sustainability. It also has to be noted that of paramount importance in using this tool is imagining the future with the view in mind of status quo and the likely outcome for sustainable development.

The second consideration is the inclusion of environmental and societal objective functions to the objective functions (economic criteria) of profit, cost, energy and yield (Edgar, et al. 2001) in optimization. The environmental objectives can be the impact or risk to environment and the societal objectives can be the impact or risk to society or common good.

These are just few examples of the needed paradigm shift in chemical engineering practice in consideration of the principle of intrinsic reality. But there can be more. A picture of the envisioned new paradigm for the practice of chemical engineering is shown in Figure 4.

Figure 4. New Paradigm in the Practice of Chemical Engineering

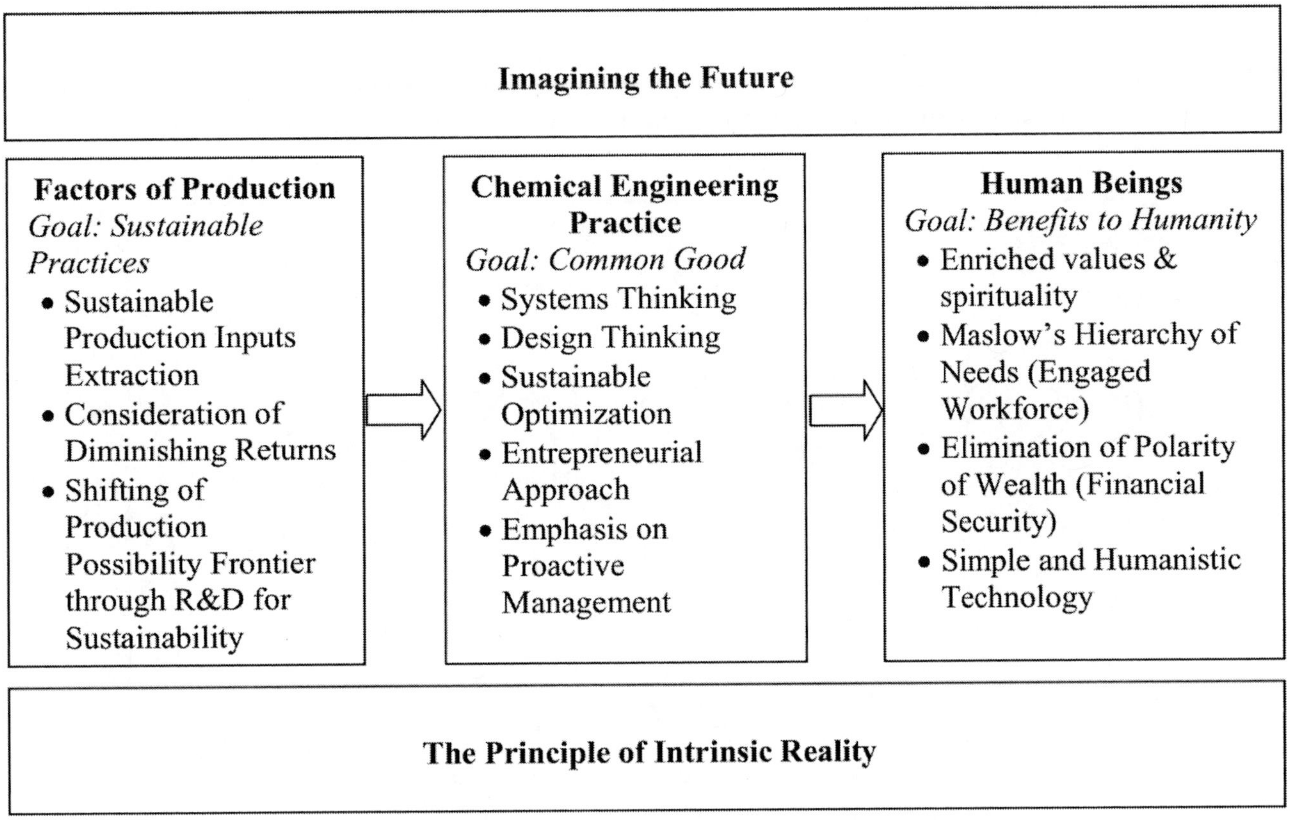

It should be realized that at the core of our economic systems is chemical engineering which can drive the economy into positive direction with the least unintended consequences. Related to that is capitalism in which upgrading of its operating systems can affect the practice of chemical engineering to a substantial degree. A discussion of this is provided in the following section.

4. Wealth Creation and Its Imperative for Sustainability

The development of this paper was inspired by the book "Freakonomics" in which the authors Steven D. Levitt and Steven J. Dubner (2005) asserted that economics can explain almost everything. It came to mind that if economics as a basic science can explain almost everything, then why not chemical engineering which was founded on chemistry, the central science?

Reflecting on it deeper, while explanations clear our doubts, we need more than that so we can tackle the world's woes. As such, this author thought that perhaps combining the two disciplines of chemical engineering and economics can be a potent force towards this end. How this can possibly be done is discussed in this section.

The starting point here is a review of the economic history of the world. In his book "A Farewell to Alms", Gregory Clark (2007) reviewed the economic history of the world in which he divided it into two periods; the period before 1800 and the period after. This is graphically represented in Figure 5.

Figure 5. Shares of Major Economies in World GDP through Time

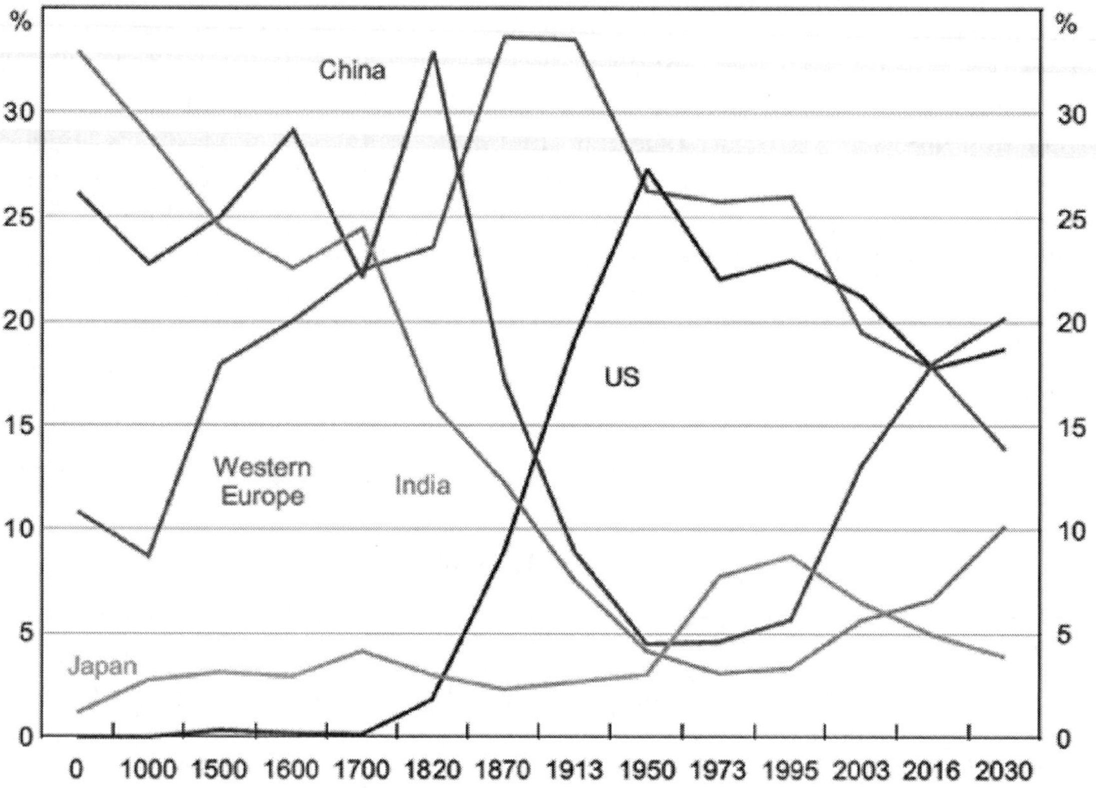

Note: Based on purchasing power parity (PPP) valuation of country GDP

Sources: International Monetary Fund (IMF), *World Economic Outlook* database, April 2011; Maddison (2006)

Thompson (2012) interpreted this graph by stating that "everything to the left of 1800 is an approximation of population distribution around the world and everything to the right of 1800 is a demonstration of productivity divergences around the world." In other words, everything to the left before Industrial Revolution depicts economic power by the sheer size of population and everything to the right after the Industrial Revolution shows the effect of mass production on country's economic output.

Clark (2007) on the other hand called the era before the industrial revolution, i.e. ca. 1800, as Malthusian era in which income per person varied across societies and epochs but growth is not observed. As Clark (2007) explained, this was due to a mechanism called Malthusian trap in which short-term increase in income through advancement in technology is dragged down by population growth. He further explicitly stated:

> "Economics as a discipline arose in the dying decades of the Malthusian era. Classical economics was a brilliantly successful description of this world. But the torrent of goods unleashed by the Industrial Revolution not only created extremes of wealth and poverty across nations, it also undermined the ability of economic theory to explain these differences."

In view of this statement, Figure 5 is reinterpreted in this paper. The contention of this author is that *polarity of wealth* can be gleaned from the graph. *Polarity of wealth* means the uneven concentration of wealth in the global economic system as some parts of the world are richer than others. The polarity can also be observed to shift as some richer parts of the world decline in economic power and some poorer parts of the world rise in economic wealth.

A historical example can be gleaned from Figure 5. India and China up until 1500 CE had 50 and 60 percent share of the world's economy, respectively. In 1 CE, China had a population of almost 60 million, while the area of the present day United States had a population of 680,000. In the late 18th century, the Industrial Revolution rendered countries with large population, just countries with the largest populations.

Looking at North America, it took them 1800 years to overtake China's economic output. However, by 1950, the US economic output was three times as great as China, even though its population is three times less. Comparing China with UK in 1913, the UK had a population of 45 million and China had 437 million but the economic output was almost the same. Analyzing what happened in Europe and later in the United States, Europe was on the lead even before 1800 (Nye 2012).

In 2010 according to data, China overtook Japan as the second biggest economy, though the numbers may not be strictly comparable. Japan's data are seasonally adjusted while those for China were not. Angus Maddison suggested that China and India were the biggest economies in the world for most of the past 2000 years. That is why economists find it a mystery why they fell so far behind and now they are flourishing ('China's economy' 2010). This may not be mysterious considering the *polarity of wealth* as was mentioned earlier and that this polarity can be argued to emanate from the Principle of Intrinsic Reality.

The reasoning here is that the world has constant mass and energy. Assuming that all these mass and energy can be harnessed to create wealth, it could mean that there is a limit for wealth creation. Theoretically, the limit can be expanded indefinitely through R&D but experience demonstrates that this was not achieved yet or getting anywhere near being achieved. As such, the constant mass and energy vis-à-vis the limit for wealth creation can be thought of as the reason for the polarity of wealth, i.e. wealth can only shift from one entity to another.

The implication of this assertion can be demonstrated by the fact that the 1% richest have accumulated more wealth than the rest of the world combined (Oxfam International 2016). The world has never seen 100% of its population to be wealthy. As such, time and again, equitable distribution of wealth is called for and mechanisms to achieve such should be made a priority in government policies.

In a more concrete way, Scharmer (2010) suggested measures to address this growing inequality which is consistent with the mindset espoused by the Principle of Intrinsic Reality. These measures are as follows:

- Creation of a New Class of Common Property Rights

The creation of common property rights is intended to address the cost to society of private economic undertakings. For example, a private company is extracting natural resources from a certain place. While the private company can profit from that endeavor, the community in that place might suffer the undesirable consequences (e.g. flooding due to forest denudation). To ensure sustainability and institutionalize the voice of future generations in this case, a trust can be established which will own the rights and distribute funds (e.g. from cap-and-trade system for carbon emissions) to citizens (one share per person), communities and the government for investment in public infrastructure.

The trust is an effective means because it does not function like a company (driven by profits) or government (driven by special interest groups) but that it is accountable for long-term sustainability of the specific commons it manage for future generations.

- Creation of a Human Right to Basic Income

One way to ensure equitable distribution of wealth is to create a human right to basic income for every human being. This right when combined with free or affordable access to healthcare and education would create a playing field where everyone has a chance to pursue aspirations and dreams. In other words, it enables individuals to put their real creativity for the benefit of larger community. Another intention of this measure is to free up paid labor (a form of dependence) so that human beings can enhance their creativity and entrepreneurial skills.

- Inclusive Capitalism

To eradicate the *polarity of wealth*, capitalism should be made inclusive so that money and capital flows are directed to all sectors of the economic system. Currently, our money system leads to accumulation of oversupply of money and capital in high financial but low environmental and social gain areas. At the same time, there is an undersupply of money and capital in areas that address societal and community needs. It implies that there is too much money in areas where it is not needed but not enough on areas where it is needed. To address this imbalance, it is recommended to redesign the system so that the current extraction-based financial system is replaced by institutions which can support the real economy and thereby serve the ecology and community.

It should be noted that shifting of wealth is also observed in various geographical regions. When this occurs, geo-political instability can result. This goes to show that if we want lasting stability and security within the limits of our intrinsic reality, then we should not tolerate unabated greed. Otherwise, *polarity of wealth* will continue to persist and we cannot realize our full potential as individuals and collaborators for common good.

5. Way Forward...

Recognizing our intrinsic reality leads us to greater understanding of our very nature bounded by the constancy of mass and energy. It shed light on why growth stabilizes after addition of more inputs. Within this purview, trade-offs are also encountered which necessitates for R&D. R&D initiatives however should be underpinned on sustainability.

Looking at the practice of chemical engineering from the lens of intrinsic reality, the approach should be futuristic while considering relevant aspects related to factors of production and human beings.

The bigger picture shows that Engineers particularly Chemical Engineers are major players in driving the real economy. As Scharmer (2010) aptly puts it, "The essence of technology and knowledge creation, and the ultimate driving force of the economy, is the same: it is the human capacity to create." As such, if sustainability is to be achieved, Chemical Engineers have a greater responsibility to innovate for regenerative and social technologies. In this way, we can stem the polarity of wealth by providing affordable technology to enable disadvantaged individuals enhance their creativity, innovation and entrepreneurship. This can help them partake in the benefits of inclusive capitalism.

As demonstrated in this paper, the synergy emanating from the convergence of Chemical Engineering and Economics can powerfully tackle the world's woes. As such, there is a good reason for chemical engineers and economists alike to collaborate particularly on sustainability initiatives.

6. References

[1] 'China's economy overtakes Japan's in real terms' 2010, *The Economist*, 16 August, viewed 5 March 2016, http://www.economist.com/node/16834943

[2] Clark, G 2007, *A Farewell to Alms: A Brief Economic History of the World,* Princeton University Press, Princeton, NJ.

[3] Edgar, T, Himmelblau D & Lasdon, L 2001, *Optimization of Chemical Processes*, McGraw-Hill, New York.

[4] Huang, Y & Wang, B 2011, 'From the Asian Miracle to an Asian Century? Economic Transformation in the 2000s and Prospects for the 2010s', in H Gerard & J Kearns (eds), *Proceedings of the Reserve Bank of Australia annual conference*, Reserve Bank of Australia, Sydney, viewed 30 January 2016,

http://www.rba.gov.au/publications/confs/2011/huang-wang.html

[5] Kolko, J 2015, 'Design Thinking Comes of Age', *Harvard Business Review*, pp. 66-71, viewed 24 February 2016, https://hbr.org/2015/09/design-thinking-comes-of-age

[6] K.N.C. 2012, 'Mis-charting economic history: more 2,000 years in a single graphic', *The Economist blog*, web log post, 20 June, viewed 29 February 2016, http://www.economist.com/blogs/graphicdetail/2012/06/mis-charting-economic-history

[7] Levitt, S & Dubner, S 2005, *Freakonomics: A Rogue Economist Explores the Hidden Side of Everything*, William Morrow, New York, NY.

[8] Malthus, T 1798, *An Essay on the Principle of Population*, J. Johnson in St. Paul's Church-yard, London.

[9] Maslow, A 1943, 'A Theory of Human Motivation', *Psychological Review*, vol. 50, pp. 370-396, viewed 24 February 2016, http://psychclassics.yorku.ca/Maslow/motivation.htm

[10] Monod, J 1949, 'The Growth of Bacterial Cultures', *Annual Review of Microbiology*, vol. 3, pp. 371-394, viewed 2 February 2016, http://www.annualreviews.org/doi/abs/10.1146/annurev.mi.03.100149.002103

[11] Nye, J 2012, 'Fascinating new graph shows the 'economic history of the world since Jesus'', *MailOnline*, 23 June, viewed 15 November 2015, http://www.dailymail.co.uk/news/article-2163610/Fascinating-new-graph-shows-economic-history-world-Jesus.html

[12] Oxfam International 2016, *An economy for the 1%, how privilege and power in the economy drive extreme inequality and how this can be stopped*, Oxford, viewed 5 March 2016, https://www.oxfam.org/sites/www.oxfam.org/files/file_attachments/bp210-economy-one-percent-tax-havens-180116-en_0.pdf

[13] Samuelson, P & Nordhaus W 2009, *Economics*, McGraw-Hill Higher Education, New York, NY.

[14] Scharmer, O 2010, 'Seven Acupuncture Points for Shifting Capitalism to Create a Regenerative Ecosystem Economy', *Oxford Leadership Journal*, vol. 1, pp. 1-21, viewed 24 February 2016, http://www.oxfordleadership.com/journal/vol1_issue3/scharmer.pdf

[15] Senge, P 2006, *The Fifth Discipline: The Art & Practice of the Learning Organization*, Doubleday/Currency, New York.

[16] Smith, R 2005, *Chemical Process Design and Integration*, John Wiley & Sons Ltd., West Sussex, England.

[17] Thompson, D 2012, 'The economic history of the last 2,000 years in 1 little graph', *The Atlantic blog*, web log post, 19 June, viewed 29 February 2016, http://www.theatlantic.com/business/archive/2012/06/the-economic-history-of-the-last-2-000-years-in-1-little-graph/258676/

Acknowledgment

The author would like to thank Hyundai Engineering and Construction Co. Ltd. for the support in this endeavor. The author is also thankful to Ms. Sofia Ayu for meticulously editing this manuscript and Mr. Marcial N. Barroga for the poster design.

445030 Heavy Metal Detection in Treated Water Via a Fabric Nanocomposite Sensor

Abstract:

We present the conception of an electrochemical sensor for on-site/portable water analysis. This flexible sensor was developed to quantify the amount of heavy metals (in real-time), alleviating the wait time, large sample size, possible contamination, and expensive analytical equipment associated with current procedures for water analysis. This sensor is composed of a multiwalled carbon nanotube (MWNT) functionalized polymer mat, for a flexible yet conductive sensor. The MWNTs are then functionalized with a cyclo-oligomeric calixarene, which is shown to selectively form a supramolecular complex with metal ions. Upon complex formation, the charge carriers are drawn away from the carbon layer and the detection is prevalent at appropriate levels. The optimization of MWNT and calixarene functionalization, as well as the selectivity and sensitivity of the sensor will be discussed.

444086 Optimization of Layered Graphene Oxide Nanostructures through Controlling the Membrane Deposition Rate Towards Size-Sieving Separation

Abstract:

Single-layer graphene oxide (GO) flakes were coated on anodized aluminum oxide (AAO) substrate through a simple vacuum filtration process, the effect of the deposition rate on the nanostructures of obtained membrane was investigated though pure water flux and organic vapor diffusion, which indicated GO membrane assembled with slow deposition rate might have more ordered lamellar structure and, correspondingly, have smaller effective pore size. In contrast, under fast deposition rate, GO flakes assembled in a more random way. Therefore, through controlling the membrane deposition rate, the nanostructure of layered GO membrane could be optimized for size dependent separation. We found after thermal reduction, 100 nm GO membrane (prepared @ 0.02 nm/s) has effective pore size smaller than 6.1 Å, which exhibit great potential in n-hexane/2,2-dimethybutane (DMB) separation.

444376 Ozone Source Apportionments of Local Anthropogenic Emissions

Abstract:

Ozone is one of six principal air pollutants designed by US EPA. NOX and VOC emissions from anthropogenic sources are the most important factors to control ozone pollutant.

In this report, photochemical dispersion models are used to estimate the effect of ozone from local anthropogenic emission sources. EKMA-type diagrams are generated for NOX- and VOC- effects. Furthermore, APCA (Anthropogenic Precursor Culpability Assessment) technology is used to estimate ozone source apportionments.

EKMA-type diagrams show that 10-30% emission reductions could reduce ozone concentrations to meet standard. Further APCA results show the average ozone contributions from regions and emission categories. APCA results explain different effects for different monitoring stations.

444142 Robust Design of a Radiant Wall Burner for a High Temperature Environment Utilizing Computational Fluid Dynamics Simulations

Abstract:

Radiant Wall Burners are often utilized in Ethylene Cracking Furnaces. The conditions and fuel gases for a specific unit can vary considerably within the unit and across similar units. With burner air preheated to high temperatures, premixing the fuel can result in dangerous and destructive flashback, while complete nozzle-mixing of the fuel can result in burner instability at startup or turn-down conditions. These challenges can be solved through the generation of a robust radiant wall burner design, which is facilitated by executing Computational Fluid Dynamics simulations of many different geometric configurations.

444397 Flare Minimization and Emission Characterization during Start-Ups of Chilling Train in Ethylene Plants

Abstract:

Flaring is a crucial safety operation to protect chemical plant personnel and equipment during plant start-up. However, it generates huge amounts of off-spec products, and causes negative environmental and social impacts, as well as results in tremendous raw material and energy losses which could be saved to generate much more needed products. Thus, cost-effective start-up flare minimization strategies which is a double-win practice, simultaneously benefiting both the industrial and environmental sustainability, are becoming more and more important and attractive to the industries.

The chilling train is a very important section for the ethylene plants and accounts for more than 50% of the total start-up time and flare emissions of the whole ethylene plants. Within this section, the charge gas from the upper stream process is refrigerated to a very low temperature (lower than -160 °C) to separate hydrogen and methane. Meanwhile, the left charge gas is liquidized for cryogenic separation. The chilling train system consumes over 40% of the refrigerating capacity of an ethylene plant. During the chilling train start-up, different operational conditions with different compositions and flowrate of precooling process will affect both product separation efficiencies, start-up stability time and flare emission.

Traditional start-up scenarios are developed based on experience and improved by trial and error, which is inefficient and dangerous. In this study, a general methodology on flare minimization for chilling train start-up operations via rigorous dynamic simulation is presented. The methodology starts with setup and validation of chilling train steady-state and dynamic simulation models. The initial state of start-up in the real plant is achieved through systematically transforming from the validated dynamic model, and several operation conditions with different compositions and flowrate of precooling stream are examined to identify an optimal initial state of start-up based on the obtained initial state. Thereafter different flare minimization strategies for the chilling train start-up are conducted on the optimal initial state model. Also, emission source characterization under different flare minimization strategies for chilling train start-up is investigated. Deep insights of the start-up strategies, emission source distribution and dynamic emission profiles are provided. This study provides detailed technical support for both the industry and environmental agencies on evaluating and developing cost-effective flare minimization strategies in the future.

439811 Purification of Ibuprofen Enantiomers By a New Technique Combining Distillation and Crystallization

Abstract:

Stripping crystallization (SC) is a new separation technique which combines distillation and crystallization. Basically, SC is operated at a triple-point condition, in which the liquid is simultaneously vaporized and crystallized due to the three-phase equilibrium. In essence, the process is continued until the liquid phase is completely eliminated and only the pure solid crystals remain in the feed. Thus, crystal washing is not required since only pure solid crystals remain in the feed and no impurities are adhered on the crystal surfaces at the end of SC. In the previous research, SC has been successfully applied to separate p-xylene from the mixed xylenes by Shiau et al. (2005, 2008, 2013).

Ibuprofen, also called 2-(isobutylphenyl)-propionic acid, is a non-steroidal anti-inflammatory drug widely used to treat headaches and minor pains. Although it is known that S-ibuprofen is responsible for the anti-inflammatory effects while R-ibuprofen might cause some long-term side effects, ibuprofen is still sold as a racemic mixture in the market due to the complexity involved in the separation of enantiomers. A thermodynamic model is developed in this work to simulate the three-phase equilibrium conditions during the SC operation. The experiments show that seeding with ultrasound mixing or magnetic stirring can be efficiently employed to purify S-ibuprofen from ibuprofen enantiomers. The experimental results, including the final enantiomeric purity and recovery ratio of S-ibuprofen, are consistent with the simulation results predicted by the model.

References:

1. Shiau, L. D.*, Wen, C. C. and Lin, B. S. “Separation and purification of p-xylene from the mixture of m-xylene and p-xylene by distillative freezing”, *Ind. Eng. Chem. Res*, **44(7)**, p2258-2265, 2005.

2. Shiau, L. D.*, Wen, C. C. and Lin, B. S. “Separation of p-xylene from the multicomponent xylene system by stripping crystallization”, *AIChE Journal*, **54(1)**, p337-342, 2008.

3. Shiau, L. D.*; Liu, K. F. "Investigations into the effects of the cooling rate on stripping crystallization", *Ind. Eng. Chem. Res*, **52**, p1716-1722, 2013.

442557 Flow Field Simulation and Distribution Performance Analysis of a Gas Pipe Distributor in Fccu

Abstract:
A fluidized catalytic cracking (FCC) is an important petroleum refinery process, which, aided by highly selective zeolite catalysts, cracks large-molecule low-value heavy oil feedstock into small-molecule highvalue
products. When spend catalyst is fed into the regenerator for reactivating, it is fluidized by the gas distributor at the bottom of regenerator. The gas distributor is designed to uniformly distribute gas to maintain the fluidized state of the fluidized bed and to provide the reaction gas to guarantee the burning coke deposited on the spend catalyst. Therefore, the gas distribution performance of the gas distributor influences the quality of initial gas distribution directly. So the pipe distributor used in a industrial FCC unit, which is composed by main pipes and a number of branch pipes with many nozzle tubes, was investigated by means of flow field calculation. The gas distribution performance of the pipe distributor was analyzed based on the simulated flow field. The simulation result shows that there are a obvious nununiform
gas distribution from the different nozzle gas flow rate. The static pressure increases gradually along the flow direction in the nozzle tube while the axial flow rate decreases, which leads to the nozzle gas flow rate increasing gradually. Furthermore, an obvious bias flow occurs at the inlet of the nozzle tube, which influences the gas flowing into the nozzle. The pressure drops performance of nozzle and pressure varying in the branch pipe are analyzed comprehensively. Due to the nonuniform
gas distribution by the gas pipe distributor in the practical usage, the problems such as overlarge temperature difference among the fluidized bed layers in the regenerator have often happened, which will not only reduce burning coke efficiency and but also decrease activity of the regenerated catalyst. Meanwhile, the erosion of the distributor also occurs, and in consequence the function of the distributor fails, which even causes the nonscheduled shutdown of the FCC unit and leads to tremendous economic loss. The support is provided for optimizations of both structural parameters and operation parameters of the gas pipe distributor and improvement of the gas distribution performance. Furthermore, improvement suggestions were provided based on the above reasoning. The results are expected to be useful for the distributor design.

441535 Hydrogen Disposal Flare System Hazard Identification and Mitigation Analysis

Abstract:

A hazard assessment on a hydrogen disposal flare system was conducted in preparation for a hydrogen turbopump test at the Air Force Research Laboratory (AFRL), Aerospace Systems Directorate at Edwards AFB, CA. The test operations required the disposal of large quantities of liquid and gaseous hydrogen through a commercial flare system to prevent the formation of an explosive environment. As part of the assessment, the primary steps in the test operation process were evaluated for the formation of a potential hazard by looking at each associated failure mode and the resulting affect on personnel and facilities. Evaluation of each operational step was conducted in a Process Failure Modes and Effects Analysis (PFMEA) that identified the probability and severity of the identified effects and associated hazards. If the probability and severity were unacceptable, mitigating steps were applied until the overall hazard risks were reduced to acceptable levels. The recommendations of this assessment were incorporated into both the flare system design and the operational procedures. This paper outlines the results of the PFMEA which include identification of the effects/hazards, available risk-mitigating techniques, and implementation of these techniques used for the mitigation of the identified hazards. Based upon the findings significant design modification to the vender's commercial equipment was needed to comply with recent changes in the code. Most of the identified hydrogen flare system hazards/effects were categorized into four groups: internal explosions, obstructions, external explosions, and maintenance degradation. Many corrective actions were implemented to mitigate both the probability and severity of the identified hazards. In addition, this paper discusses the installation of oxygen monitoring systems on the flare system, determination of appropriate system purge requirements, installation of purge system status alarms, incorporation of a triple redundant flare ignition system with status indicators, and a number of other hazard risk mitigation techniques. This comprehensive description of the hazards and mitigation of a hydrogen disposal system is a plant is vital part to safe operation of both hydrogen commercial and aerospace facilities.
<align="center">

444020 Thermodynamic Modeling of Aqueous Lithium Sulfate and Its Solubility in the Presence of Na+, K+ and Mg2+ Sulfates

Abstract:

This work is an extension of the previous works done by our group towards development of a rigorous and comprehensive thermodynamic model for high salinity produced water that is essential for mass and heat balance calculations and simulation and design of novel treatment processes. Specifically this work presents a thermodynamic model that accurately represents all thermodynamic properties of aqueous lithium sulfate solution including the solubility of Li_2SO_4 in water and also in the presence of Na^+–K^+–Mg^{2+} sulfates. By explicit consideration of solution chemistry of each subsystem, symmetric electrolyte NRTL model is used to correlate available experimental data of all pertinent thermodynamic properties including activity coefficient, osmotic coefficient, excess enthalpy, and liquid heat capacity over the entire concentration range. This facilitates quantification of eNRTL binary molecule–molecule, molecule–electrolyte, and electrolyte–electrolyte interaction parameters. Also, solubility data are used to regress the thermodynamic constants required for modeling the liquid–solid equilibrium of lithium sulfate in water and in the presence of Na^+–K^+–Mg^{2+} sulfates. Displaying a significant prediction ability, the model accurately represents the thermodynamic behavior of the system from infinite dilution up to saturation and temperatures from 298.15 K to 573.15 K.

442068 Enhanced Furfural Production By an Innovative Bottom Dividing Wall Column with a Decanter Configuration

Abstract:

Enhanced Furfural Production by an Innovative Bottom Dividing Wall Column with a Decanter Configuration

Nguyen Van Duc Long, Le Cao Nhien, Le Quang Minh and Moonyong Lee†

School of Chemical Engineering, Yeungnam University, Gyeongsan 712-749, South Korea

Furfural, which are used to derive other industrial chemicals, has been identified as one of the major bio-based platform chemicals that can compete with petroleum-based chemicals. However, the current commercial furfural process has low yield and requires intensive energy. This paper reports the simulation, detailed design and optimization of an industrial process recovering furfural from lignocellulosic biomass, a renewable and inexpensive feedstock. The aqueous feed mixture from the acidic hydrolysis of biomass was fed to the separation process to recover furfural as the main product, along with methanol as valuable byproducts. The purification process was intensified by applying an innovative bottom dividing wall column with a decanter configuration (BDWC-D). The response surface methodology (RSM), which allows the interactions between variables to be identified and quantified, was used to optimize the BDWC-D structure. The predictions by the RSM showed satisfactory agreement with the rigorous simulation results. The heat pump technology and heat integration were then employed to improve the purification process. The results showed that the proposed sequence could save the overall energy requirements and total annual cost significantly as compared to the conventional sequence. This sequence can be employed for both grass-root and retrofit designs.

Keywords: Lignocellulosic Biomass; Furfural; Energy Efficiency Improvement; Dividing Wall Column; Response Surface Methodology

Acknowledgments

The authors wish to acknowledge the financial support from the R&D Convergence Program of NST (National Research Council of Science & Technology) of Republic of Korea and KITECH (Korea Institute of Industrial Technology) (ES150001).

*** Formatting Issue - Best Available Paper/Graphic ***

444278 Insight into the Mechanism of Toxic Chromium (VI) from Wastewater Using Magnetite Coated PINE Powder

Abstract:

INSIGHT INTO THE MECHANISM OF TOXIC CHROMIUM (VI) FROM WASTEWATER USING MAGNETITE COATED PINE POWDER

Pholosi A., Ofomaja A., Naidoo E.B.

Biosorption and Wastewater Treatment Research Laboratory, Department of Chemistry,

Faculty of Applied and Computer Sciences, Vaal University of Technology, P. Bag X021, Vanderbiljpark, 1900 South Africa.

INTRODUCTION

A knowledge of the interaction between pollutant molecules or ions and active sites on an adsorbent surface is an important phenomenon in many industrial processes such as catalysis, adsorption and other separation techniques. The use of lignocellulosic materials for Cr(VI) adsorption has gained interest over the years **(Miretzky and Cirelli, 2010)**. The adsorption properties of lignocellulosic materials has been attributed to acidic functional groups on their surfaces which can take part in ion-exchange, complexation, chelation and hydrogen bonding with pollutant species in solution **(Vithanage et al., 2015)**. Two mechanisms for lignocellulosic adsorption of Cr(VI) have been proposed; (i) an electrostatic mechanism in which at low pH, the protonated acidic functional groups on the biomaterial carrying a positive charge attracts the negatively charged Cr(VI) species **(Demirbas, 2005)** and (ii) the adsorption-coupled reduction mechanism in which Cr(VI) is reduced to Cr(III) either in solution (Direct reduction) or after being adsorbed (Indirect reduction) by electron-donor groups of the biomaterial that have lower reduction potential values than that of Cr(VI) **(Gao et al., 2008)**. The Cr(III) ions formed in solution may then form complexes with Cr(III)-binding groups on the biomaterial surface or released into bulk solution by repulsion between Cr(III) and other positively charged groups on the biomaterial **(Miretzky and Cirelli, 2010)**.

METHODS

Sodium hydroxide treated pine coated with magnetite (NTP-NC) was prepared by co-precipitation method. The effect of solution pH was performed in a batch system in which 0.5 g of the adsorbent was contacted in six 250 cm^3 beakers containing 75 cm^3 of 75 mg/dm^3 Cr(VI) solutions set at pH 1, 2, 4, 6, 8 and 10 at 26 °C and agitated at 200 rpm for 2 h and the clear solution analysed for Cr(V/I) left using UV-Vis spectrophotometry at 460 nm

RESULTS

The results of the effect of initial solution pH on the adsorption of Cr(VI) onto NTP-NC is shown in Fig. 1. To confirm the operating mechanism, the change in $[H^+]$ concentration at each initial pH, the oxidation/reduction potential before and after adsorption and the amounts of Total Cr, Cr(VI)and Cr(III) present in solution at each pH was monitored for NTP-NC and the results displayed in Fig1 . The change in $[H^+]$ concentration ($\Delta[H^+] = [H^+]_{final} - [H^+]_{initial}$) at the end of the adsorption process for each initial solution pH was recorded and plotted against the initial solution pH. The results showed that the change in $[H^+]$ concentrations at the end of the adsorption process where higher for lower initial solution pH's (where lower amounts of Cr was found in solution) than for higher initial solution pH's (where higher amounts of Cr was found in solution) and the change in $[H^+]$ concentration became small and

*** Formatting Issue - Best Available Paper/Graphic ***

almost constant after initial solution pH 4. This is consistent with the assumption that the removal of Cr from solution is associated with [H^+] ion consumption and that as the concentration of [H^+] ions reduces in solution, the Cr removal reduces. Cr(VI) ion in acidic solution has a positive redox potential value (above 1.3 V at standard conditions) and in the presence of an electron donor becomes unstable and can be reduced to Cr(III) **(Park et al., 2008)**. The results show that the initial ORP values (before adsorption) for the Cr solutions at pH's of 1, 2, 4, 6, 8, and 10 were 339, 280, 165, 55, -53 and -172 mV for NTP-NC respectively. The positive values of ORP at lower solution pH's indicates higher possibility for reducing Cr(VI), while the lower and negative ORP values indicates lower possibility for reducing Cr(VI). On addition of the adsorbents into the solutions, a change in ORP were observed . The values of ORP at the end of the adsorption process became 330, 268, 30, 9 -29 and -61 for NTP-NC respectively. A reduction in ORP at the end of the adsorption stage signifies that reduction has occurred in the system. **Yu et al. (2014)** showed the amount of Cr(VI) removed from aqueous solution in contact with zero-valent iron (nZVI) could be related to the change in ORP and pH of the solution. The authors observed a decrease in ORP from 400 to 100 mV and a corresponding 10 unit pH increase as Cr(VI) was removed from solution.

Finally, analysis of the prepared Cr solutions at the various solution pH's showed that the concentration of Total Cr obtained from ICP and that of Cr(VI) using UV-vis measurements were very similar, suggesting that Total Cr was almost equal to Cr(VI) in solution. After contacting the various solutions with the adsorbents, the concentration of Total Cr in the pH 1 solution was observed to have reduced to 21.56 mg/dm^3 for NTP-NC, while the Cr(VI) concentrations was 10.66 mg/dm^3 for NTP-NC. With solution pH 2, the concentration Cr(III) left in solution was lower than that of solution pH 1 for both samples. The reason is that at pH 1 there is an abundance of H^+ ions in solution which accumulates on the functional groups on the adsorbent making the surfaces of the adsorbent positively charged and repelling away the positively charged Cr(III) ions from the adsorbent surfaces **(Suksabye et al., 2009)**. At solution pH 2, it is believed that there is less accumulation of H^+ ions at the adsorbent surfaces which led to reduced electrostatic repulsion and increased adsorption of Cr(III) from solution. This increase in Cr(III) adsorption is also confirmed by the reduction in Total Cr in solution between solution pH 1 and 2. The concentrations of Cr(VI) left in solution at pH 2 was higher than at pH 1 for NTP while for NTP-NC the Cr(VI) concentrations were lower at pH 2. For solution pH's higher than 2 and up to 10, the residual Cr(III) in solution becomes very low and almost constant. This is consistent with the small change in [H^+] in solution which is almost constant at this pH range.

FTIR Evidence for the adsorption of Cr(III)

When Cr(III) is formed in solution via adsorption-coupled reduction mechanism it may undergo hydrolysis and/or complexation depending on the solution pH and the total Cr(III) concentration. Hydrolysis may lead to the formation of various Cr(III) species with an increase in solution pH which gives rise to a reduction in its solubility and increasing the chances of precipitation of the Cr(III) species. At solution pH 1 – 4, Cr^{3+} and $CrOH^{2+}$ are the major species which can be adsorbed by negatively charged ligands on the biomaterial surface. On the other hand, at solution pH range of 4 – 10, Cr(III) species (hydroxo complex), such as $CrOH^{2+}$, and $Cr(OH)_3$, are the dominant forms in the aqueous system and at high pH is readily transformed into soluble complex, . When NTP-NC was contacted with Cr(VI) solution for 2 hr, the final solution pH was increased to pH 2.53. The FTIR spectra for NTP-NC (Fig. 2) shows the presence of characteristic Fe-O peaks of bulk Fe_3O_4 at 553.26 cm^{-1}, 564.36 cm^{-1}, 570.21 cm^{-1} and 576 21 cm^{-1}. Functional groups associated with the pine includes the broad band at 3250.99 cm^{-1} representing –OH of cellulose, 2905.17 cm^{-1} representing C-H stretching, while the peaks at 1627.69 and 1602.87 cm^{-1} shows the COO^- and C=O bonds respectively. After 2 hr contact with Cr(VI) solution, the peak at 564.34 cm^{-1} was observed to have split into several new peaks (Fig. 3b (ii)). These emerging peaks are characteristic of Fe-O bonds of maghemite (γ-Fe_3O_4), for example the characteristics peaks of maghemite was observed at 595.21, 685.65 cm^{-1} and 582.70, 634.91 cm^{-1} suggesting a change in the iron oxide phase (Fe_3O_4 to γ-Fe_2O_3) due to the conversion of Cr(VI) to Cr(III) by magnetite nanoparticles. According to **Peterson et al. (1997)**, Cr(III)(hydr)oxide produced is precipitated on the magnetite surface. In this study, α-Cr_2O_3 peak were observed at 604.63 cm^{-1} indicating that Cr(III) was formed during the adsorption process. Finally it was also observed that the carboxylate group at 1635.55 cm^{-1} was shifted to 1637.73 cm^{-1} while a new peak was observed at 1737.72 cm^{-1} suggesting that complexation between Cr(III) and Carboxylate ions also occurred.

REFERENCE

Demirbas, A. Energy Source 27 (2005) 1449–1455.

Miretzky, P., Cerelli, A.F. A review. J. Hard. Mater. 180 (2010) 1-9.

Vithanage, M., Rajapaksha, A.U., Ahmad, M., Uchimiya, M., Dou, X., Alessi, D.S., Ok, Y.S. J. Environ. Manage. 151 (2015) 443-449.

D. Park, C.K. Ahn, Y.M. Kim, Y.-S. Yun, J.M. Park, J. Hazard. Mater. 160 (2008) 422–427.

P. Suksabye, A. Nakajima, P. Thiravetyan, Y. Baba, W. Nakbanpote.. J. Hazar. Mater. 161 (2009) 1103–1108.

R.F., Yu, F.H., Chi, W.P., Cheng, J.C. Chem. Eng. J. 255 (2014) 568–576

M.L. Peterson, A.F. White, G.E. Brown, G.A. Parks Environ. Sci. Technol. 31 (1997) 1573–1576.

Fig. 1 Effect of solution pH on the adsorption of Cr(VI)

*** Formatting Issue - Best Available Paper/Graphic ***

Fig, 2: FTIR of NTP-NC

443101 Thermodynamic Modeling of HNO_3-H_2so_4-H_2o Ternary System with Symmetric Electrolyte NRTL Model

Abstract:

Nitric acid concentration/sulfuric acid concentration (NAC/SAC) process has been widely used for concentrating dilute aqueous nitric acid and recovering spent sulfuric acid. Nitric acid is concentrated by employing sulfuric acid to bind water and break nitric acid-water azeotrope at approximately 68 wt. % nitric acid. To support heat and mass balance calculations and process simulation for NAC/SAC processes, we develop a comprehensive thermodynamic model for nitric acid-sulfuric acid-water ternary system based on previously published thermodynamic models of nitric acid-water [Wang et al., Fluid Phase Equilibria, (2015) (DOI: 10.1016/j.fluid.2015.04.015)] binary and sulfuric acid-water [Que et al., Journal of Chemical & Engineering Data, 56 (2011) 963-9772] binary systems developed with symmetric electrolyte NRTL (eNRTL) activity coefficient model. The present work takes into account one-step hydration of nitric acid molecules, two-step hydration of sulfuric acid molecules, and further hydration of hydronium ions. With up to three temperature coefficients for each eNRTL binary interaction parameter and up to two temperature coefficients for each chemical equilibrium constant, the ternary model is validated with boiling point data from 383.15 to 471.45 K and activities at 273 K. Furthermore, contour plots of boiling points, vapor phase composition, and specific heat capacity of the ternary system are generated for engineering use.

Keywords:

Nitric acid, Sulfuric acid, Electrolyte NRTL model, Vapor-liquid equilibrium, Boiling point, Heat capacity, Nitric acid concentration/sulfuric acid concentration (NAC/SAC)

444551 Continuously Adjustable, Molecular-Sieving "Gate" on 5A Zeolite for Distinguishing Small Organic Molecules By Size

Abstract:

Zeolites/molecular sieves with uniform, molecular-sized pores are important for many adsorption-based separation processes. Pore size gaps, however, exist in the current zeolite family. This leads to a great challenge of separating molecules with size differences at ~0.01 nm level. Here, we report a novel concept, pore misalignment, to form a continuously adjustable, molecular-sieving "gate" at the 5A zeolite pore entrance without sacrificing the internal capacity. Misalignment of the micropores of the alumina coating with the 5A zeolite pores was related with and facilely adjusted by the coating thickness. For the first time, organic molecules with sub-0.01 nm size differences were effectively distinguished via appropriate misalignment. This novel concept may have great potential to fill the pore size gaps of the zeolite family and realize size-selective adsorption separation.

444791 Dissolution and Precipitation of Minerals Under Different Alkali Concentrations in a Reactive CO_2 Batch System

Abstract:

The primary focus of this research is to study the dissolution and precipitation of minerals and salts from Berea sandstone and brine samples taken from the Great Salt Lake in a reactive CO_2 batch system. Various experiments were constructed to measure dissolution and precipitation reaction rates under different alkali concentrations. Changing the alkali concentrations of the Great Salt Lake brine samples produced changes in dissolution and precipitation of minerals from the Berea sandstone and the Great Salt Lake brine samples. Data from ICP-MS was used to quantify mineralogical changes in the samples and Micro-CT was used to measure the porosity change in the Berea sandstone samples. The results from these analytical instruments showed patterns of mineralogical dissolution and precipitation in the post-experimental fluid that correlate to the change in alkali concentrations.

*** Formatting Issue - Best Available Paper/Graphic ***

443365 Technical Reconstruction of Double Effect Distillation for Methanol Process

Abstract:

Abstract: Methanol is one of important basis raw materials of chemical industry and also an excellent environment-friendly fuel, which make it as a significant commodity in the global market. Distillation is an essential unit operation for purified production of methanol. Product quality, production cost, environmental impact and energy consumption are largely related to distillation process.

Our research effort, under this circumstance, has been devoted since 2002 to energetically optimize 4-column methanol distillation system that is widely adopted in methanol units from coal, with reduced cycling water requirement. This patented process technology saw its industrial application in Chinese largest single unit of 1800Kt/a of Baofeng Group located at North China, with the lowest steam consumption of 0.9/1 t(steam)/t(methanol) compared with normally 25% higher for both simulative case study and on-site operation. Methanol product is in accordance with O-M-232G and the recycling water is reduced significantly from 67 t(CW)/t(methanol) to 52 t(CW)/t(methanol).

The flow diagram of this advanced double effect distillation process is showed in Fig.1. Its advantages arise for limitations found in current units: (1) heat exchange network failed to fully recover energy causing large CW flow to remove extra heat; (2) for pressured column, bubble point reflux need larger reflux ratio than the sub-cooled; (3) the heat of distillate from pressured column was normally ignored; (4) the low concentration of methanol of feed inlet to recovery column leads to higher reflux radio, responsible for unsatisfied unit energy efficiency.

Fig.1 The double effect distillation process.

(Column1.Light ends column; Column 2.Pressured column;

Column 3.Atmospheric column; Column4.Recovery column)

As a solution, a new flow diagram is built up and showed in Fig.2. (1) HEN totally reorganized (not completely show); (2) sub-cooled reflux to pressured column caused 10% reduction to the reflux ratio, economizing steam consumption; (3) top product of pressured column give heat to crude methanol before feeding light ends column; (4) change product and by-product position of recovery column to give efficient unit energy consumption.

Fig.2 The reconstruction of distillation process.

(Column1.Light ends column; Column 2.Pressured column;

Column 3.Atmospheric column; Column4.Recovery column)

It is estimated that the production of methanol will be over 60 million tons in 2014. If this reconstruction could be applied in the whole methanol industry, 1.3 million tons of standard coal would have been saved. This is equivalent to reduce 5 million tons emission of carbon dioxide, so the environmental impact can be largely decreased. Field operation shows reconstruction perfect substitution worth to spread to the whole methanol industry, energetically efficient and low environmentally impacting.

444353 A Combined Rheology and Stability Study on Emulsified Acid for High Temperature Applications

Abstract:

HCl emulsion has been a popular formulation to stimulate carbonate reservoirs due to the various advantages that emulsified HCl has over ordinary HCl; including but not limited to deeper penetration and less corrosion damages. Emulsified HCl has even more potential advantages if utilized properly. The reaction kinetics, stability and rheology of emulsified HCl plays a significant role on achieving successful acidizing treatment.

The objective of this paper is to experimentally study the factors affecting the stability of emulsified HCl. The emulsion was prepared using 15wt% HCl and diesel with cationic emulsifier. The acid solution-to-diesel ratio was 70:30. The effect of stirring intensity, concentration of the emulsifier, addition of corrosion inhibitor, temperature, addition rate of acid phase to oil phase and mixing time of the emulsion have been studied. Stirring intensity was varied from 600 to 1500 rpm. Emulsifier dosage ranged between 5 to 20 gpt. Temperature used for heating, to mimic reservoir condition, was as high as 120°C.

The results showed that there is an optimum condition for emulsifier dosage and for mixing time of the emulsion. The optimum emulsifier dosage was found to be 10 gpt with 5 minutes of mixing time. As for the addition rate of acid phase to oil phase, 1.4 ml/min of drop-wise addition was found to give a very stable emulsion. The effect of stirring intensity on the emulsion stability was proportional, the higher the rpm speed the better the stability. Corrosion inhibitor, on the other hand, played a role as a destabilizer. With regards to heating temperature, it was found out that the higher the temperature, the faster the emulsion breaks.

The novelty of this study is the comprehensive approach in combining stablity and rheology of emulsified acid for high temperature applications. The particle size analysis and solution morphology help in better understanding HCl emulsion and how it works.

443503 A Computational Analysis of Fire Scene for Safety in a Compartment

Abstract:

This article details numerical prediction of fire spread carried out to study the distribution of carbon monoxide concentration, smoke dispersion, soot concentration and temperature in laboratory compartment, using a commercial fire modeling software tool. The heat release rate curve predicted from the simulation was compared with literature values carried out in a road tunnel. The time variations of carbon monoxide transportation, smoke concentration and temperature spread in compartment room are reported to evaluate the effect on humans. The soot concentration was described and used throughout this study as an indicator of safety in the compartment room. The effects of vent dimension and ventilation capacity were analyzed by smoke detector response times. This work provides improvements derived from reinterpretation of actual fire scene considering the ability of mechanical exhaust system and ventilation in compartment room to mitigate the hazard of fire.

444359 Development of Novel Emulsified Acid Using Waste Oil

Abstract:

In carbonate reservoirs, hydrochloric acid (HCl) is commonly used for stimulation purposes. The main disadvantage for using HCl is its high and rapid reactivity allowing for less control on acid reaction with formation. In addition, HCl may react with the tubing causing severe corrosion and damages. Several slow-reacting acid systems, though HCl-based, have been developed to retard the HCl reaction and control its reactivity to allow deeper penetration in carbonate reservoirs before it breaks out and reacts with formation rock. However, the selection of the hydrocarbon phase of emulsified acid is critical to the creation and stability of the emulsified acid system. Diesel oil, crude oil and solvents such as xylene have been reported to be successful in the field.

In this work, alternative and less expensive source of oil is developed to replace the current practices. Emulsified acid is a water-in-oil emulsion where oil is the continuous phase and acid is the dispersed one. Industrial and filtered waste oil was used to prepare the emulsified acid of this work. Extensive work has been carried out to study the chemical composition of the waste oil, namely SARA and to select the proper emulsifier. Emulsions were prepared using 15% HCl with emulsifier concentration ranging from 0.5 to 2.0 vol% and 70:30 of acid-to-oil ratio. Droplet size distributions were measured and used to benchmark with other emulsions. Emphasis was given to the stability of the emulsion especially at high temperatures. In order to apprehend the effect of corrosion inhibitor on the emulsion stability, its concentration was varied (up to 0.3 vol%). Later, rheological analyses of the waste oil emulsion were compared to the commonly used diesel-based emulsified acid.

Results showed the potential use of waste oil to prepare high temperature emulsified acid (up to 100°C). The addition of corrosion inhibitor to the emulsified acid caused a negative effect on stability resulting in a less stable emulsion. On the other hand, addition of cationic emulsifier helped to formulate a more stable emulsion. Apart from that, the optimum concentration of the emulsifier was at 2.0 vol%. The rheological data confirms the stability of the new emulsified formulation at higher temperatures. This paper summarizes the findings of using waste oil emulsion and recommends it for field applications. The newly proposed emulsified acid introduces waste oil as a replacement of diesel.

The novelty of using waste oil will help to significantly reduce the expenses of formulating emulsified acid. Since waste oil is basically waste it has a good impact on the environment, that being said, instead of throwing it, the industry will be able to recycle it.

443764 Optimization of Heat Exchanger Network in Steam Assisted Gravity Drainage Oil Sands Surface Facilities

Abstract:

Steam assisted gravity drainage (SAGD) is the most viable and environmentally safe recovery technology for extracting heavy oil and bitumen. It is a thermal in-situ production process where parallel wells are drilled horizontally into an underground bitumen reservoir. Steam is produced at the central processing facility (CPF) and injected into the reservoir through the wells. The steam heats the bitumen to a point where gravity allows it to flow down to the lower well where the mixture of bitumen and water is then pumped to the surface. The water and bitumen are separated at the CPF. Typically, more than 90% of the energy consumed in producing a barrel of bitumen using SAGD is used to generate the steam that is injected into the bitumen reservoir. Fossil fuels is employed in the process, which causes GHG (greenhouse gas) emissions. The usage of hot and cold utility is reduced by increased heat exchange and optimization of the heat exchanger network, which results in reduction of GHG emissions. In this work, pinch analysis is performed to minimize energy consumption.

443855 Modeling the Thermal Behaviors of an Ultracapacitor Module

Abstract:

Ultracapacitors, also known as supercapacitors, have the potential to meet the increasing power requirements of the sustainable energy systems for the grid-scale energy storage systems and electrified vehicles. As compared to batteries, ultracapacitors offer a higher power density, higher efficiency, and longer shelf and cycle life. Because the performance of an ultracapacitor depends on temperature, it is important to calculate accurately the thermal behavior of a single ultracapacitor cell and the module composed of multiple ultracapacitor cells for the efficient and reliable systems integration of an ultracapacitor in grid and transport applications.

In this work, modeling is performed to study the effect of the layout configugation of the ultracapacitor cells on the thermal behaviors of an ultracapacitor module. The ultracapacitor module is subject to the charge/discharge cycling with constant-current between the nominal voltage of the ultracapacitor module and the half of this voltage. The charge/discharge current values examined are 50, 100, 150, and 200 A. The validation of the modeling approach is provided through the comparison of the modeling results with the experimental measurements.

443856 Modeling the Electrolyte Flow Behaviors of a Zinc/Bromine Flow Battery Module

Abstract:

The zinc/bromine (Zn/Br_2) flow battery is a promising rechargeable power source for energy storage systems because of its inherent chemical simplicity, high degree of reversibility at the electrodes, good energy density, and abundant low-cost materials. The Zn/Br_2 flow battery system, intended for the grid-scale energy storage system, is to be made in modular form. A Zn/Br_2 flow battery module is composed of the multiple stacks consisting of bipolar carbon electrodes. Since the stacks require electrolyte circulation on both the negative and positive sides, two plumbing loops are provided. The flow loops contain an electrolyte storage tank and a circulation pump. A heat exchanger may be included for temperature control in the flow loop. The design of the electrolyte circulation plumbing determines the distributions of electrolyte flow and pressure within the stacks and the efficiency of the stack is dependent on the flow distribution throughout the stack. The modeling of the distributions of electrolyte flow and pressure within the module is needed to optimize the plumbing for electrolyte circulation and thereby to improve the performance of the module. In this work, modeling is performed to study the effect of the layout configuration of the plumbing for electrolyte circulation on the distributions of electrolyte flow and pressure within a 50 kWh Zn/Br_2 flow battery module composed of four stacks. It is seen that the plumbing layout for electrolyte circulation can have a significant effect on the flow rate distribution of electrolyte through the stacks.

443974 Observer-Based Model Predictive Control for Time-Delay Systems

Abstract:

Model predictive control is one of the most widely used advanced process control technologies in industry.[1] It has many capabilities such as dealing with input/state constrains and uncertainties. Observer-based model predictive control was developed to estimate states information when the states are not measurable directly. Many chemical processes have time delays in either states or measurements, such as polymer production and some bioprocesses. [2] In this presentation, we are proposing an unknown input observer based model predictive control algorithm for time-delay systems. Unknown input observer is an observer design to eliminate certain types of unknown inputs in the system. Stability of unknown input observer-based model predictive control without time delay was provided by our previous study. Terminal cost and terminal region are designed for a class of non-linear time delay systems. The close-loop stability of the controller is provided. [3][4]

An example of bioreactor is provided which has uncertain time delays in measurements but with a known upper bound. Two measurements are provided which are online biomass and offline ethanol measurements. Model-plant mismatch is considered as unknown inputs in the system. State estimation and control are calculated using the observer and an optimization algorithm.

Reference:

[1] Marcus Reble et al., Model predictive control of constrained non-linear time-delay systems. IMA Journal of Mathematical Control and Information, 2010, 1-19

[2] Seung Cheol Jeong and PooGyeon Park, Constrained MPC algorithm for uncertainty time-varying systems with state delay, IEEE TRANSACTIONS ON AUTOMATIC CONTROL, 50, 2005, 257-262

[3] Mayuresh Kothare et al., Robust constrained model predictive control using linear matrix inequalities. Automatica, 32, 1996, 1361-1379

[4] Wook Hyun Kwon et al., A simple receding horizon control for state delayed systems and its stability criterion, Journal of Process Control, 13, 2003, 539-551

444044 Organic Solar Cells

Abstract:

This paper describes highlights the need of next-generation "organic solar cells". There is a continuing need for organic solar cells and arrays with very high specific power densities (1000-5000 kW/kg) for generating power in a new generation of ultra light space payloads example in NASA. An emerging technology with promise to meet these ambitious goals are organic solar cells based on very thin films of discrete layers. Theoretical efficiencies for multi junction solar cells are more ~70% as compared to ~30% for single junction devices. There is continuous development going on high mobility, band gap engineered organic donors and acceptors, as well as development of three junction devices and scaled up devices on lightweight flexible polymer substrates. The main interest in this paper is organic semi-conducting crystal: Pentacene, which has photo-conducting states. Organic solar cells manufactured today have low efficiency other than their stability issues. The efficiency of organic solar cells can be improved by either designing polymer materials that absorbs more of solar spectrum or by minimizing the charge collection efficiency by controlling the morphology of polymer films. Once fully developed, organic solar cells would revolutionary power the industry due to their flexibility and minimal weight

444055 A Refined Non-Random Two-Liquid Segment Activity Coefficient Model for Solubility Modeling

Abstract:

Accurate correlation and robust prediction for complex molecule solubility or solid-liquid equilibrium (SLE) in solvents and solvent mixtures have drawn much interest recently [Tanveer et al., *Chem. Eng. Prog.* **2014**, 110 (9), 37-47]. Solubility models such as the Non-Random Two Liquid Segment Activity Coefficient (NRTL-SAC) model [Chen and Song, *Ind. Eng. Chem. Res.* **2004**, 43, 8354-8362] are now routinely used in the industries. However, an important unfilled need is to model the solubility in oligomers/polymers/surfactants such as asphaltene nanoaggregates in petroleum fluids and drug molecules in lipid-based or solid dispersion formulation [Tung, *Org. Proc. Res. Dev.* **2013**, 17, 445-454]. This paper presents a refinement to the very successful NRTL-SAC model to better account for the contribution due to size differences of molecules. Specifically, the Flory-Huggins combinatorial entropy expression for conceptual segments in the model is substituted with the Staverman-Guggenheim expression for real molecules. We present the rationale for the refined model, the updated conceptual segment numbers for common solvents, and the newly identified conceptual segment numbers for common drug excipients. The model performance is illustrated by prediction of drug molecule solubility in various lipid solvents and solvent mixtures, compared to experimental data and previous work by other models. The refined NRTL-SAC model shows enhanced capability to correlate and predict drug molecule solubility in common solvents and excipients used in drug formulation.

443640 Preparation and Characterization of (*FURCRAEA spp*) as Green Support of Iron Nanostructured Catalyst

Abstract:

PREPARATION AND CHARACTERIZATION OF (*FURCRAEA spp*) AS "GREEN" SUPPORT OF IRON NANOSTRUCTURED CATALYST

Karen G. Bastidas G[1], Anamaria Barrera B[1], Cesar A. Sierra A[2], Hugo R. Zea R[3]

[1] Department of Chemical and Environmental Engineering. National University of Colombia (Bogota).

[2] Associate professor. Director of the Research Group of Macromolecules. National University of Colombia (Bogota).

[3] Associate professor. Director of the Research Group of Materials, Catalysis and Environment. National University of Colombia (Bogotá).

Abstract.

The availability of natural water resources has declined over time, becoming one of the biggest problems today. Environmental risk in Colombia's rivers has increase due to the widespread use of harmful chemicals in a variety of processes, ranging from illegal mineral extraction to industry wastewater, e.g. Medellin's river, due to its strategic location along a textile industrial corridor might be subject to dye contamination; the gold and platinum mining zones in Choco and Antioquia regions are being heavily affected by the use of mercury in the extraction processes. In 2009, 520 tons of mercury were imported by miners working in this area, amount that exceeds several times the mercury used by regulated industry in previous years (Ministry of Mines and Energy of Colombia, 2007). Not only the environmental risk is a concern but also the economic conditions of many low-income families whose livelihood in many cases depends exclusively of the of the mining.

Due to the issues raised above, the main goal of this study is to research, develop and implement innovative technologies in the synthesis of highly efficient, low cost and environmentally friendly iron based catalysts capable of degradate pollutants, ensuring operational stability. A native fiber (Fique, FURCRAEA spp) has been selected as a solid support for these catalysts due to its abundance, low cost and special physical and chemical characteristics.

The combination of cellulosic fibers with iron oxide nanoparticles could provide exceptional biodegradable composite materials for the treatment of organic dyes and heavy metals present in wastewater. Likewise, the use of iron particles immobilized on fique fibers allows an easy separation catalyst-treated water, turning the biocatalyst in highly recyclable and reusable material.

Fique fiber was initially characterized by determining the percentage of ashes, total lignin, soluble and insoluble lignin, cellulose, hemicellulose, total solids and humidity, obtaining 0.94, 22.58, 0.5, 22.8, 36.3, 27.5 0.701 and 2.1%, respectively. Additionally, the isoelectric point (pzc), acidic and basic sites were measured.

Then, the fiber was subjected to a pretreatment in order to expose more active sites which on time serve as nucleation points for the deposition of nanoparticles. This was done by means of an ultrasound-assisted procedure, varying the sonication time (30, 60 and 90 minutes). The effect of the sonication was followed by measuring the contents of lignin, cellulose and hemicellulose within the fiber after each time. 60 minutes of sonication was found to be most beneficial time since the amount of lignin and hemicellulose present in the fiber was reduced, while conversely the amount of crystalline cellulose exposed increases.

In order to improve the chemical ability of the nucleation sites to interact with iron and then to form stable and size controllable nanoparticles, a cationization procedure was performed on the fiber; this was achieved by the exposure of the fiber to acidic and basic conditions promoting cationization. Three cationization times (1, 2 and 3 hours) were studied and their effect on the fiber chemical surface was evaluated using FTIR-ATR and XRD analysis. The results showed 3 hours as the optimal cationizing time.

Subsequent to the cationization of the fiber surface, a procedure of ionic/cationic exchange was implemented by exposing the fiber to a solution of iron salts in order to replace the cations in the fiber surface and anchoring the iron nanoparticles. Different conditions were studied in order to synthetize zero valent iron nanoparticles (nZVI) and others with different oxidation states. Finally, to track the impregnation process on the fiber, samples from the precursor solution were taken at different times and its iron concentration was determined by atomic absorption spectrophotometry (AAS). Complementary characterization performed by X-ray diffraction, BET surface area and electronic microscopy, among others.

Key words: Natural support (fique), iron nanoparticles, bionanocatalyst.

444057 Explaining the Paradox of Asphaltene Precipitation with n-Alkanes

Abstract:

Explaining the paradox of asphaltene precipitation with n-alkanes

Md Rashedul Islam and Chau-Chyun Chen
Department of Chemical Engineering, Texas Tech University

Abstract
Paraffins, specifically alkanes, promotes asphaltene precipitation when they are blended with asphaltene-containing petroleum fluids. The alkane/oil volume ratio required for onset of asphaltene precipitation increases with the carbon number and yields a maxima. Experimental study indicates that the maxima lays at C7[1] or between C8 -- C10 [2]. However, precipitation yield decreases asymptotically with increase of the carbon number even if excessively high alkane/oil volume ratio is used [3]. For example, n-pentane be a better solvent for asphaltene than n-hexane in regard of both onset of precipitation and precipitation yield. However, high carbon number alkane beyond the maxima, n-hexadecane for instance, can be a better solvent at onset of precipitation point and a poorer solvent in regard of precipitation yield than n-pentadecane. A thermodynamic explanation for this paradox is necessarily important to find the compatibility of bitumen or crude oil with paraffinic hydrocarbons. In this work, this paradox is explained by the aggregation thermodynamics of asphaltene precipitation [4]. Both the enthalpy and entropy of mixing are investigated to explain the effect of paraffin carbon number on asphaltene precipitation. Athabasca and cold lake bitumen samples are studied.

References
[1] G. Hotier and M. Robin, "Action de divers diluants sur les produits pÃ©troliers lourds: mesure, interprÃ©tation et prÃ©vision de la floculation des asphalténes," Oil & Gas Science and Technology, vol. 38, pp. 101-120, 1983.
[2] I. A. Wiehe, H. W. Yarranton, K. Akbarzadeh, P. M. Rahimi, and A. Teclemariam, "The paradox of asphaltene precipitation with normal paraffins," Energy & fuels, vol. 19, pp. 1261-1267, 2005.
[3] D. L. Mitchell and J. G. Speight, "The solubility of asphaltenes in hydrocarbon solvents," Fuel, vol. 52, pp. 149-152, 1973.
[4] M. Wang, Y. Hao, M. R. Islam, and C. C. Chen, "Aggregation Thermodynamics for Asphaltene Precipitation," AIChE Journal, Submitted.

Using a Large Data Bank to Develop a Robust and Simple Model to Estimate the Saturation Pressure of Oil Reservoirs

By

Hassan Alkandari[1], Adel Elsharkawy[1], Adel Malallah, Osamah Alomair

Petroleum Engineering Department, College of Engineering & Petroleum
Kuwait University
Po Box 5969 Safat 13060-Kuwait

Abstract

Measurements of saturation pressure are crucial for all hydrocarbon reservoir fluids. Below the crude oil saturation pressure gas reaches a critical saturation; then, a two phase flow occurs and results in decreasing oil production and recovery. To maximize oil production and recovery, the reservoir pressure has to be maintained closer to the original saturation pressure. This pressure is normally measured using bottom-hole samples or surface recombined samples of oil and gas. Occasionally, the samples become unavailable and the pressure needs to be estimated using computational methods.

In this study, a large data bank of 231 crude oil compositions and saturation pressure measurements including literature data, unpublished data, and newly measured data were used to develop two empirical models to predict the saturation pressure of a variety of crude oils if the oil sample is unavailable or the experimental measurements are unreliable. The first proposed model utilizes the extended compositions of hydrocarbons up to the heptane plus fraction in addition to non-hydrocarbons. The second model uses the lumped compositions of light, intermediate, and heavy components in addition to non-hydrocarbon components as the input variables. The lumped model has the advantage of using fewer input parameters while maintaining the thermodynamics. The accuracy and validity of both models to calculate the saturation pressure for volatile oils, black oils, and heavy oils are presented using several compositional data. The models performance is also compared to the Peng-Robinson equation-of-state (PR-EOS), and the Soave-Redlich-Kwong equation-of-state (SRK-EOS) as well as all published methods that use compositions as input variables. The comparison indicates that the proposed models are much simpler and more accurate than the other computational methods. The proposed models treat the heptane-plus fraction as a single component, thus eliminating the splitting and characterizing the plus fractions and the binary interaction parameters needed for the EOS's calculations.

Key words: Saturation pressure (Ps), Bubble-point pressure (BPP), Equation of state (EOS), Correlations, and PVT Data

[1] Corresponding Authors: Hassan Alkandari , Hassan@hkandari.com, Adel Elsharkawy, adel.elsharkawy@ku.edu.kw

Introduction:

Reservoir fluid properties are an essential source of information for all calculations on hydrocarbon reserves. Accurate estimation of the pressure-volume-temperature (PVT) measurement is crucial. Of all the PVT properties, saturation pressure (Ps) is considered the most important parameter. It is well known that gas starts to flow once the reservoir pressure drops below the bubble point pressure and the gas saturation reaches a critical value. To minimize the gas flow and maximize the oil recovery, it is very important to maintain the reservoir pressure close to the original bubble point pressure.

Although the PVT properties are experimentally measured in laboratories and provide reliable results, this process is expensive, time consuming, and sometimes unavailable; therefore, measured field data (separator gas oil ratio, separator pressure, stock-tank oil specific gravity, and reservoir temperature) or reservoir fluid compositions are used to estimate the bubble point pressure.

There are two methods in the petroleum industry to calculate the saturation pressure of crude oil when experimental is unavailable; either using production data or compositional data.

Methods Based on Production Data (Composition is unavailable)

Numerous attempts have been made since the 1940s to develop correlations to estimate the saturation pressure from production data, such as producing gas oil ratio, oil gravity, gas gravity, reservoir pressure and temperature. Table 1 presents summary of all the methods that use production data to estimate the saturation pressure.

Methods based on Compositional data

Many authors have reported that the application of any correlation in other regions results in serious errors. These conclusions arise from the fact that regional crude oils represent a physical-chemical trend; therefore, all attempts to develop a general correlation using field measured parameters to estimate the saturation pressure from a large data bank have failed. For this reason, Elsharkawy (2003) made the first attempt to correlate the saturation pressure to the reservoir fluid composition instead of using the production data. In recent years, there have been several attempts to estimate the saturation pressure of oil reservoirs. These methods are reported in Table 1.

Elsharkawy (2003) presented a correlation of the bubble point pressure for the oil reservoir using the same input data needed for the equation-of-state (EOS) calculations, i.e., the compositional data and the reservoir temperature. Elsharkawy's model is based on the saturation pressure measurements of 60 crude oil samples from the Middle East and 75 collected from the literature. He demonstrates that the calculation of the saturation pressure depends on the type of EOS, the number of pseudo components, the method used to characterize the pseudo-components and the binary interaction number. Elsharkawy compared the performance of his model to the SRK-EOS and PR-EOS for black oils, volatile oils, heavy oils, and synthetic oils. He indicated that his model eliminates splitting the heavy plus fraction, characterizing the pseudo-components and inclusion of the binary interaction parameters needed for the EOS calculations.

Later, Tale et al. (2009) made the second attempt to model the saturation pressure of crude oil using 132 data samples published by Elsharkawy (2003). Tale's model, which is based on the alternating conditional expectation (ACE), has a set of 14 equations with 42 constants. The average absolute relative error (AARE) reported by Tale et al. indicates that the model did not perform well with the selected data. The ACE model has an AARE of 12.3% compared with AARE of 6.7% by Elsharkawy (2003) and 8.1% AARE for the SRK-EOS.

AlQuraishi (2009) attempted to model the saturation pressure using Elsharkawy's (2003) data. The model was developed using linear genetic programming. He correlated the saturation pressure to the formation temperature, methane content, and heptanes plus molecular weight. Thus, AlQuraishi's model neglected the effect of the intermediate and heptane plus content on the saturation pressure. The error analysis reported by AlQuraishi indicates that his model has a 5.8% error compared to 7.7% by Elsharkawy, 9.9% using the SRK-EOS, and 10.1% using the PR-EOS.

Bandyopadhyay and Sharma (2011) presented a model to predict the bubble point pressure, using temperature and fluid composition. They used 129 experimental and literature data originally published by Elsharkawy (2003) to develop their model. The model introduced a temperature interaction parameter that depends on compositions. The proposed model has a set of 24 coefficients and the estimated bubble point pressure is calculated as the sum of the series calculations. They reported an average absolute error (AAR) of 7.83% using their model compared to 8.3% using Elsharkawy's simple model (2003), 10.15% using SRK-EOS, and 10.58% using the PR-EOS.

Khamehchi et al. (2011) used an intelligent proxy simulator (IPS), which is based on an adaptive neural fuzzy inference system (ANFIS) and a structured artificial neural network (ANN) to predict the bubble point pressure and asphaltene onset pressure. The model uses a reservoir fluid composition and temperature to predict the Ps. It is not clear how the Ps is calculated from the proposed method, and they did not present the accuracy of the proposed method compared to the other published methods.

Farasat et al. (2013) presented a new mathematical approach to calculate the crude oil saturation pressure as a function of the temperature and the reservoir fluid composition. The model developed by Farasat et al used 130 experimental data, which were originally published by Elsharkawy (2003). Farasat used a support vector machine (SVM) to model the saturation pressure. They did not clearly demonstrate how to calculate the saturation pressure from the SVM model. Moreover, they did not indicate the overall average absolute deviation (AAD) for their model compared to other published models and both the PR-EOS and SRK-EOS for the 130 data.

Ahmadi et al. (2014) also presented an ANN technique called gene expression programming (GEP) to determine the bubble point pressure of the oil samples. Similar to Farasat et al. (2013), Ahmadi's model used most of the data previously published by Elsharkawy (2003) to construct a non-linear equation using reservoir fluid compositions and temperatures as input parameters. The proposed model is highly complicated and consists of 8 equations that have several constants compared to the one simple equation proposed by Elsharkawy (2003). They compared the performance of their proposed model to the SRK-EOS and PR-EOS; however, their paper does not mention how the heptanes plus fraction were treated and characterized for the EOS calculations. They reported an AAD of 4% using the proposed model compared to 8% using Elsharkawy's model (2003), 10.9% by PR-EOS, 10.5% by SRK-EOS, and 15% by Bandyopadhyay and Sharma (2011).

Gholami et al. (2014) estimated the bubble point pressure using a support vector regression SVR and a supervised learning algorithm SLT based on the statistical learning theory, and then the ACE was used. The model used reservoir fluid compositions and temperature to estimate the bubble point pressure. Gholami et al. used the same data originally published by Elsharkawy (2003). They presented three attempts based on genetic programming to correlate the bubble point pressure: SVR, ACE, and power-law committee machine (PLCM). The ACE-based method resulted in a set of three equations with a matrix of [8 by14] constants. Gholami et al. compared the accuracy of their model to the SRK-EOS and PR-EOS predictions reported by Elsharkawy (2003). Based on their AARD analysis, the PLCM model has the lowest errors followed by the SVR model, Elsharkawy's model, ACE model, PR-EOS, and SRK-EOS.

Lately, Jarrahian et al (2015) presented two empirical models to estimate the saturation pressure of black oils. The first model is based on production information. The second model is based on compositional information mostly from Elsharkawy (2003) and Heidaryan and Moghadasi (2011). They have not presented comparison of the accuracy of their model to Farasat (2013) model, Ahmadi (2014) model, and Gholami (2014) model.

This paper is concerned with the estimation of saturation pressure using compositional data. Therefore, the objective is to develop a robust yet simple model to estimate the saturation pressure for various types of reservoir oils using a larger data bank than that used by other published studies. The accuracy of the proposed models is compared to previously published models as well as SRK-EOS and PR-EOS.

Data bank

This study uses a large data set of 231 crude oil compositions and saturation pressure measurements. The data set consists of 130 data sets originally reported by Elsharkawy (2003); in addition to others collected from literature (Jacopy, 1958; Vogel, 1980; Coats, 1986; Droham, 1980; Dansh, 1992; Moharam, 1996; Rosenegger, 1999); 94 new data sets that have not been utilized before; as well as 7 newly measured samples at the Petroleum Fluid Research Center (PFRC) in Kuwait University. These data represents a wide range of compositions, such as volatile oils, black oils, and heavy oils. The data cover a pressure range of 537 to 5065 psia and a temperature range of 58 to 319 °F, crude oil compositions from C_1 to C_{7+}, non-hydrocarbons, such as N_2, H_2S, and CO_2, and the molecular weight and the specific gravity of the heptane plus fraction.

Data screening and classification

The compositions and the saturation pressure measurements in the data bank are subjected to a meticulous screening process to check the validity of the data. Our screening criteria uses the following steps: the sum of the molar compositions of all the components must add up to one, otherwise, the sample is eliminated. Samples with a missing molecular weight or specific gravity of the plus fraction are also eliminated.

After filtering, the data were reduced to 214 and divided into 179 measurements used to develop the new models, and 35 data sets are used for the blind test of the robustness of the proposed models. Statistical analysis of the input variables of the data set used in training and testing the extended and the lumped models are illustrated in Tables 2.

Several of the reported compositional data have extended analysis up to (C_{20+}). The following techniques are used for lumping or regrouping the extended analysis into single pseudo-fractions (C_{7+}). The mole fraction (X_i) of the individual components of the extended analysis is equal to the plus fraction (Ahmad, 2011).

$$X_{C_{7+}} = \sum_{i=7}^{n} X_i \qquad i=7, 8 \ldots n \qquad (1)$$

The average molecular weight (Mw) of the individual component of the extended analysis is equal to the molecular weight of the plus fraction.

$$Mw_{C_{7+}} = \sum_{i=7}^{n} \frac{X_i Mw_i}{X_{C_{7+}}} \qquad i=7, 8 \ldots n \qquad (2)$$

The density of the total plus-fraction can be calculated as follows:

$$\gamma_{C_{7+}} = \frac{X_{C_{7+}} Mw_{C_{7+}}}{\sum_{i=7}^{n} \frac{X_i Mw_i}{\gamma_i}} \qquad i=7, 8 \ldots n \qquad (3)$$

It is important to note the significance of the density of plus fraction because its density determines its paraffinic naphthenic aromatic (PNA) content.

Equation-of-state calculations

The EOS calculation of the saturation pressure is compared with the newly developed models as well as the other methods. The two EOS calculations considered in this paper are the ones most considered and widely used in the petroleum industry to evaluate the volumetric and phase behavior of reservoir fluid properties: the Soave-Redlich-Kwong EOS (SRK-EOS) and the Peng-Robinson EOS (PR-EOS). It should be noted that the accuracy of any EOS calculation of the saturation pressure largely depends on the number of pseudo-components and the splitting schemes used to divide the heptane plus fraction; as well as the characterization and calculation of the critical properties of the pseudo-components and the binary interaction parameters. Therefore, the two EOS calculations of the saturation pressure are not the same, even though the number of pseudo-components, characterization of the pseudo-components, and binary interaction parameters were the same.

Model development

As previously stated, the objective of this work is to provide the petroleum reservoir engineer with a simple yet accurate tool to estimate the saturation pressure for a variety of crude oils using compositional data when experimental is unavailable. Two models are presented in this study to calculate the saturation pressure of the crude oil samples. The first model uses all input parameters, such as the molar compositions of the hydrocarbon and non-hydrocarbon components similar to the EOS models. This model is referred to as the extended model. The second model uses fewer input parameters in which some of the hydrocarbon components have been lumped into a single group; this is designated as the lumped model. Both models are simpler than all published correlations, physically correct, and have an acceptable accuracy compared with the well-known EOS.

Figure 1 shows a correlation matrix of all the data available under consideration. This figure also indicates the relative importance of each of the input variables into the first proposed model. It agrees with most published correlations and the EOS that the methane content, heptane plus content and temperatures have the greatest impact on the saturation pressure followed by the intermediate content. The data understudy indicates that the non-hydrocarbon has the smallest impact. This result is simply a reflection of the molar contribution of each component comprising the crude oil composition.

Model I - Extended Model

The extended model uses 13 variables comprising the molar compositions of the pure components, heptane plus properties, and reservoir temperatures, which are similar to the input parameters for the EOS. Figure 2 shows the correlation coefficient graph of all the input parameters in the extended model. Thus, the elimination of the intermediate effect has a negative impact on the accuracy of the predictive model. Our proposed extended model has the following form:

$$Ps = 124.72N_2 + 17.57CO_2 + 22.55H_2S + 64.22C_1 - 9.80C_2 - 52.49C_3 + 6.16C_4 - 19.22C_5 - 23.63C_6$$

$$- 21.43C_{7+} + 435.31\gamma_{C7+} + 1.14Mw_{C7+} + 4.29T \quad (4)$$

Where Ps is the saturation pressure in psia, T is the temperature in °F, and the hydrocarbon and non-hydrocarbon compositions are expressed in mole.

Model II - Lumped Model

The second model is much simpler and uses only 7 -variables (the lumped model). Several lumping schemes are adapted in the petroleum industry for phase behavior modeling and reservoir simulation of the petroleum reservoir fluids (Ahmed, 2011). Hydrocarbon components are grouped into a pseudo fraction based on their molecular weight (Whitson, 1980), the similarity of their physical-chemical properties (Lee, 1979), their carbon number, their true boiling point or their number of families based on the chemical structure (Becker et al., 2015). (Emera and Samara, 2005) proposed lumping the hydrocarbon components into volatile (C_1 & N_2), intermediate (CO_2, H_2S, C_2 to C_4) and heavy categories (C_{5+}) to calculate the minimum miscibility pressure. (Duan et al., 2013) proposed lumping several components into a single carbon number (SCN) with predetermined values for properties, such as critical pressure, temperature, and eccentric factor. Others have proposed optimal lumping schemes based on the objectives (Lin et al., 2008). In lumped model, $C_{1+}N_2$, C_2, CO_2, and H_2S, the intermediate (C_3 to C_6), heptanes plus (C_{7+}) and its properties, and temperature are the input variable. The correlation matrixes of the lumped components are provided in Figure 3. Even though the non-hydrocarbon has the least effect their contribution is important, especially the H_2S in the case of sour crudes. Both models presented by AlQuraishi (2009) and Bandyopadhyay and Sharma (2011) have eliminated the effect of H_2S.

The proposed model has the following form:

$$Ps= 82.115*(C_1+N2)-11.635*C_2+39.158*CO_2+ 38.244*H_2S -1.217*(C_3—C_6) - 890.701*(C_{7+}/MW_{7+}*\gamma_{7+}) + 4.217 *T-1042 \quad (5)$$

Where Ps is the saturation pressure in psia; and T is the temperature in °F.

Result and discussion

Model accuracy

The accuracy of the two proposed models used to calculate the saturation pressures of the previously described crude oil samples is shown in Table 3. This table shows reports the average relative error, average absolute relative error standard deviation and the correlation coefficients for all the methods considered in this study. Table 3 also reports the accuracy of the PR-EOS, SRK-EOS, Elsharkawy's model, AlQuraishi's model, and Bandoypandyay and Sharma's model and Jarrahian model. The methods proposed by Gholami et al., Ahmadi et al., and Jarrahian et al are included in Table 3, even though they produce a unreasonable estimate of the saturation pressures, as explained in the appendix.

This table indicates that the extended and the lumped models have an (AARE) in the order of 8% and 8.5%, respectively. It should be noted that the complications in the Bandoypandyay and Sharma model did not improve the accuracy of the saturation pressure, because it had a large error. It is also noted that Jarrahian et al model produces an average error of 35%. Compared with the EOS, the lumped model has a few constants and eliminates the splitting, characterizing the pseudo-components and binary interaction parameters necessary for the EOS calculations. The lumped model as well as the extended model are very simple and require one-step calculations compared with all the other methods reported in Table 3.

A cross-plot of the calculated saturation pressure using the extended model based on the training data versus the measured data is illustrated in Figure 4A. This figure indicates that most of the data falls on the unit slope line, and there is no over-fit. This trend is supported by the high correlation of regression of the model (97.6%) presented in Table 3. Figure 4B shows a cross-plot of the testing data (28 from literature and 7 new measured in PFRC at Kuwait University). The extended model indicates an excellent agreement with all of the 28 data sets chosen at random, with the exception of a few data points at higher pressures. Additionally, the model has a perfect match for the 7 newly measured samples.

Figure 5A shows the performance of the lumped training-model for capturing the measured saturation pressure for most of the 179 samples. A cross-plot of the calculated saturation pressure using the lumped model based on the training data versus measured data is depicted in Figure 5B. This figure indicates that most of the data scatters falls to the unit slope line. The lumped model also has a high coefficient of regression of 97.4%, which is indicated in Table 3.

Models Validity

The accuracy and validity of the models for estimating the saturation pressure of various types of crude oil are reported in Table 4. This table illustrates the capability of the newly proposed models to predict the saturation pressure for volatile oils, black oils, and heavy oils. Table 4; provides the compositions, properties of the heptane plus fraction, temperature, and measured saturation pressures of the samples. Additionally, the calculated saturation pressure using the PR-EOS, SRK-EOS, Elsharkawy's model (2003), AlQuraishi's model (2009), and Bandoypandhyay and Sharma's model (2011), Jarrahian et al (2015); as well as the extended and lumped models are provided in Table 4. It is clear from this table that the proposed models have a wide range of applications regardless of crude oil types. We have also included information regarding the 7 newly measured data; at the PFRC, compositional, and saturation pressure. At the bottom of Table 4, we have provided the unpublished data of the compositions; and the saturation pressures at low temperatures. It is well known that the saturation pressure of a crude sample is measured at low temperatures in addition to the reservoir temperature as a method to check the quality of the sample.

Figure 6A reports the error of all the methods considered in this paper to predict the saturation pressure for volatile oils (heptane plus content below 30%). Both the extended and the lumped models have an AARE of nearly 4% followed by Bandoypandhyay and Sharma's model (2011); and Elsharkawy's model (2003); the SRK-EOS has nearly twice the AARE of the newly proposed model. AlQuraishi's model (2009) has an AARE of 12% because the model is based on methane and neglects the effect of heavy and intermediate components, whose contribution is highly significant in the reservoir and production performance of volatile crudes. Jarrhaine et al (2015) model is not suitable for volatile oils. It shows a 35% error in estimated saturation pressure.

The performance of the extended and lumped models compared to the various methods for the black oil samples ($35< C_{7+} <55\%$) is provided in Figure 6B. The newly proposed models have an AARE close to 10%; followed by Bandoypandyay and Sharma's model 14%, the SRK-EOS and PR-EOS 16%, and Elsharkawy 17%. AlQuraishi's model has an AARE of 24%. Jarrhain et al (2015) model shows a 33% error in estimated saturation pressure even though the authors claim that the model is suitable for black oils.

Figure 6C shows the error level of all the methods as well as the proposed models for the newly measured crude oil samples at Kuwait University using the PFRC. Again, Jarrhain et al (2015) model show a 40% error in estimated saturation pressure.

The performance of all published models at low temperature is illustrated in Figure 6D. The low-temperature calculation indicates that Bandoypandhyay and Sharma's model (2011) has accuracy problems and is not valid at low temperatures because the authors proposed a temperature-composition related coefficient. Also, Jarrhain et al (2015) model is not suitable at low temperature. It shows a 40% error in estimated saturation pressure.

The ability of the proposed models to capture the thermodynamic is thoroughly checked against its independent variables. The validity of the model is compared to the experimentally measured data as well as the predictions by other methods used to capture the physics of increasing the saturation pressure as a function of temperature is illustrated in Figure 7. Two main points are observed here; first, the model matches the measured saturation pressures much better than the other methods. Second, it follows the physical trend predicted by the EOS's. Figure 8 shows the model performance for capturing the effect of changing the saturation pressure as a function of the molecular weight of the heptane plus fraction compared with the SRK-EOS and PR-EOS. The extended model follows the trend of the PR-EOS in increasing the saturation pressure. Elsharkawy's model (2003) was unable to predict the correct trend as predicted by the EOS's. The model sensitivity in capturing the change in the saturation pressure as a function of increasing the methane content is shown in Figure 9. The proposed model as well predicts the correct trend of increasing the saturation pressure when increasing the methane content.

Conclusion

In this paper, two models are presented to estimate the saturation pressure of a variety of crude oils when measures are unreliable or samples are unavailable. The two models are developed using compositional data and saturation pressure measurements from a large data bank comprising literature data, unpublished data, and newly measured and reported data of 231 crude oil samples. The input parameters in the first model are the extended hydrocarbon and non-hydrocarbon molar percentages and reservoir temperatures. The second model uses lumped hydrocarbon fractions of intermediate, and heptane plus fractions in addition to its properties; and the reservoir temperature as input parameters. The accuracy of the two models has been tested and compared to that of the SRK-EOS and the PR-EOS; as well as published correlations that use compositional data as the input parameters. The comparison indicates that the two models are much more accurate than all the published methods. The validity of both models has also been tested using various types of crude oils samples and compared to experimentally measured saturation pressures and estimates using the EOS as well as the published correlations. The proposed models are physically correct and much simpler than all the published methods. The models do not require tedious computations and eliminate splitting of the plus fraction, characterization of the pseudo-components; as well as the binary interaction parameters that are required for the EOS calculations.

Acknowledgement

The authors thank the General Facility Lab at Kuwait University for their cooperation and conducting the compositional and saturation pressure measurements of the 7 new samples described in this paper using Petroleum Fluid Research Center (PFRC) project no. GE 01107.

References

Agarwal, R.K., Li, Y.K., Nghiem, L., 1990. A regression technique with dynamic parameter selection for phase – behavior matching. SPE Res. Eng. February, 115 – 120.

Ahmadi, M. A., Zendehboudi, S, James, L., Elkamel, A., Dusseault, M., Chatsiz, I., Lohi, A., 2014, New Tools to determine Bubble Point Pressure of Crude Oils: Experimental and Modeling Study, *Journal of Petroleum Science and Engineering, v.123, 204-216.*

Ahmed, T., 2011, Splitting and lumping schemes of the plus fraction, working guide to vapour-liquid phase equilibria calculations, Chapter 9, Hydrocarbon Phase Behaviour, Gulf Publishing Co. 111-126.

Al-Gathe, A., Abd-el Fattah, Kh., El-Mtwally, K., 2014, New artificial intelligent approach for bubble point pressure, Int. J innovation and Sci. Research, Vol.4, No.2, July, 121-135.

Al-Marhoun, M.A. 1985. Pressure-Volume-Temperature Correlations for Saudi Crude Oils. Presented at the SPE Middle East Oil Technical Conference and Exhibition, Bahrain, 11–14 March. SPE-13718.

Almehaideb, R.A. 1992. Improved PVT Correlations for UAE Crude Oils. Presented at the Middle East Oil Show and Conference, Bahrain, 15-18 March. SPE-37691.

Al-Najjar, H.S., Al-Soof, N.B.A., and Al-Khalisy, K.M. 1988. Correlations for Bubble-Point Pressures, Gas Oil Ratios and Formation Volume Factors for Iraqi Crude Oils. *Journal of Petroleum Research* (June 1988): 13.

AlQuraishi, A. A., 2009, Determination of crude oil saturation pressure using linear genetic programing, Energy & Fuel, 23, 884-887.

Al-Shammasi, A.A. 2001. A Review of Bubblepoint Pressure and Oil Formation Volume Factor Correlations. *SPE Res. Eval. & Eng.* 4 (2): 146-160. SPE71302.

Asgarpour, S., McLauchlin, L., Wong, D. et al. 1989. Pressure-Volume-Temperature Correlations for Western Canadian Gases and Oils. *J Can Pet Tech.* 28 (4): 103.

Asoodeh, M., Bagheripour, P.,2012, Estimation of bubble point pressure from *PVT* data using a power-law committee with intelligent systems, *J. Pet. Sci. and Eng., Vol. 90–91, July 2012, Pages 1-11.*

Bandyopadhyay, P., Sharma, A., 2011, Development of a new semi analytical model for prediction of bubble point pressure of crude oils, *Journal of Petroleum Science and Engineering, Volume 78, Issues 3–4, September 2011, Pages 719-731.*

Becker, P. J., Celse, B., Guillaume, D., Dulot, H., Costa, V., 2015, Hydrotreatment modelling for variety of VGD feedstocks: a continuous lumping approach, Fuel, 139, 133-143.

Coats, K.H., and Smart, G.T., 1986, Application of regression base EOS PVT Program to laboratory data, SPERE, May 277-299.

Dansh, A. Xu, D. H., Todd, A. C., 1992 A grouping method to optimize reservoir description for compositional simulation of gas injection process, SPERE, Aug. 343-348.

De Ghetto, G. and Villa, M. 1994. Reliability Analysis on PVT Correlations. SPE-28904, Presented at the European Petroleum Conference, London, United Kingdom, 25-27 October.

Dindoruk, B. and Christman, P.G. 2001. PVT Properties and Viscosity Correlations for Gulf of Mexico Oils. Presented at the SPE Annual Technical Conference and Exhibition, New Orleans, 30 September-3 October. SPE-71633

Dokla, M.E. and Osman, M.E. 1992. Correlation of PVT Properties for UAE (United Arab Emirates) Crudes. *SPE Form Eval.* (March 1992): 41.

Droham, J.K., 1988, Enhancing the evaluation of PVT data, OSEA paper presented the 7th Offshore South East Asia, Feb. 2-5, Singapore.

Duan, J., Wang, W., Liu, H., Cong, J., 2013, modelling the characterization of the plus fraction by using continuous distribution function, Fluid Phase Equilibria, 345, 1-10.

Elam, F.M. 1957. *Prediction of Bubble Point Pressures and Formation Volume Factors from Field Data.* MS thesis, University of Texas at Austin, Austin, Texas.

Elmabrouk, S., Zekri, A., Shirif E., 2014, The prediction of bubble point pressure and bubble point oil formation volume factor in absence of PVT analysis, PET. Sci. & Tech., 32, 1168-1174.

Elsharkawy, A. M., 2003, An empirical model for estimating the saturation pressures of crude oils, Journal of Petroleum Science and Engineering 38 (2003) 55-77.

Elsharkawy, A.M. and Alikhan, A.A. 1997. Correlations for predicting solution gas/oil ratio, oil formation volume factor, and undersaturated oil compressibility. *J. Pet. Sci. Eng.* 17 (3–4): 291-302.

Emera, M., Samara, H., 2005, Use of genetic algorithm to estimate CO2-oil minimum miscibility pressure- a key parameter in design of CO2 miscible flood, J. Pet. Sci. Eng., Vol. 46, issue 1-2, 37-52.

Farasat, A., Shokrollahi, A., Arabloo, M., Gharagheizi, F., Mohammadi, A., 2013, toward an intelligent approach for determination of saturation pressure of crude oil, Fuel Processing Technology 115 (2013) 201-214.

Frashad, F., LeBlanc, J.L., Garber, J.D. et al. 1996. Empirical PVT Correlations for Colombian Crude Oils. SPE-36105 Presented at the SPE Latin American and Caribbean Petroleum Engineering Conference, Port of Spain, Trinidad and Tobago, 23–26 April. -MS.

Gharbi, R.B.C., Elsharkawy, A.M., 1999, neural network model for estimating the PVT properties of Meddle East crude oils, SPE Reservoir Eval. Eng. 2 (3), 255-265.

Gholami, A., Asoodeh, M., Bagheripour, P., 2014 How committee machine with SVR and ACE estimates bubble point pressure of crudes, *Fluid Phase Equilibria, Volume 382, 25 November 2014, Pages 139-149.*

Glasø, Ø. 1980, Generalized Pressure-Volume-Temperature Correlations. *J Pet Tech.* 32 (5): 785-795.
Heidaryan, E., Moghadasi, J., 2011, Crude saturation pressure through imperial correlation, Thermodynamics 2011, Athens, 31 August- 3 September 2011.
Hasan, M., 1993, PVT Correlations for Indonesian Crude Oils. *Proc.* 22nd Annual Convention of the Indonesian Petroleum Association, Indonesia, Vol.1, IPA93-2.1-070.
Jarrhian, A., Moghadsi, J., Heidaryan, E., 2015, Empirical estimating of black oil bubblepoint (saturation) pressure, J. Petroleum Science and engineering, (126), 69-77.
Jacopy R.H., Berry V.J.,1958, A method for predicting the pressure maintenance for reservoir producing volatile oil, AIME 213, 59-65.
Kartoatmodjo, T., Schmidt, Z. 1994, Large Data Bank Improves Crude Physical Property Correlations. *Oil Gas J.* 92 (27): 51–55.
Kartoatmodjo, T.R.S. and Schmidt, Z. 1991, *New Correlations for Crude Oil Physical Properties*, SPE 23556.
Khairy, M., El-Tayeb, S., Hamdallah, M., 1998. PVT Correlations Developed for Egyptian Crudes. *Oil Gas J.* 96 (18): 114.
Khamekhcki, E., Behvandi, R., Rashid, F., 2011, Prediction of bubble point pressure & asphaltene onset pressure during CO2 injection using ANN & ANFIS models, Pet. Sci. & Tech. Vol. 1 No.2, 35-45.
Labedi, R.M. 1982, *PVT Correlations of the African Crudes*, PhD thesis, CSM.
Labedi, R.M. 1990, Use of Production Data to Estimate the Saturation Pressure, Solution GOR, and Chemical Composition of Reservoir Fluids, SPE-21164-MS. Presented at the SPE Latin America Petroleum Engineering Conference, Rio de Janeiro, Brazil, 14-19 October,
Lasater, J.A. 1958, Bubble Point Pressure Correlations. *J Pet. Tech.*10 (5): 65–67. SPE-957-G.
Levitan, L.L. and Murtha, M. 1999. New Correlations Estimate Pb, FVF. *Oil Gas J.* 97 (10): 70.
Lee, S., 1979, Experimental and theoretical studies on the fluid properties required for simulation of thermal process, Paper SPE 8393.
Li, Y.K., Nghiem, L.X., Siu, A., 1985. Phase behavior computations for reservoir fluids: effect of pseudo-components on the phase diagrams and simulation results. J. Can. Pet. Technol. Nov. – Dec., 29 – 36
Lin, B., Leibovici, C. F., Jorgensen, S. B., 2008, optimal component lumping: problem formulation and solution techniques, Computer & Chem. Eng., 32, 1167-2272.
Macary, S.M. and El-Batanoney, M.H. 1992, Derivation of PVT Correlations for the Gulf of Suez Crude Oils. *Proc. of the* 11th EGPC Pet. Exp. & Prod. Conference, Vol. 1, 374, Cairo, Egypt,
Moharam, H.M., 1995. Prediction of viscosity of heavy petroleum fractions and crude oil using a corresponding states method. Ind. Eng. Chem. Res. 34, 4140 – 4144.
Moradi, B., Malekzadeh, E., Amani, M., Boukadi, F.H., Kharat, R., 2101, Bubblepoint pressure imperial correlation, SPE 132756.
Obomanu, D.A. and Okpobiri, G.A. 1987, Correlating the PVT Properties of Nigerian Crudes. *J. Energy Res. Tech.* 109 (4): 214-217.
Okoduwa, I. G., Ikiensikimama, S.S., 2010, Bubble point pressure correlation for Niger Delta Crude Oils, SPE 136968.
Omar, M.I. and Todd, A.C. 1993. Development of New Modified Black Oil Correlations for Malaysian Crudes. SPE-25338 Presented at the SPE Asia Pacific Oil and Gas Conference, Singapore, 8-10 February.
Owolabi, O.O. 1984. *Reservoir Fluid Properties of Alaskan Crudes*. MS thesis, University of Alaska, Fairbanks, Alaska (May 1984).

Pedersen, K.S., Thomassen, P., Fredenslund, Aa., 1988a. Characterization of gas condensate mixtures. AIChE Spring National Meeting, March 6 – 10, New Orleans, LA.

Peng, D. y., Robinson. D. B., 1976. A new two-constant equation of state , Ind. Eng. Chem. Fund. 15, 59-63.

Petrosky, G.E. Jr. 1990. *PVT Correlations for Gulf of Mexico Crude Oils*. MS thesis, University of Southwestern Louisiana, Lafayette, Louisiana.

Petrosky, G.E. Jr. and Farshad, F. 1998. Pressure-Volume-Temperature Correlations for Gulf of Mexico Crude Oils. *SPE Res Eval & Eng.*1 (5): 416-420. SPE-51395.

Riemens, W.G., Schutle, A.M., Jong, L.N.J., 1988. Birba field PVT variations along the hydrocarbon column and confirmatory field tests. J. Pet. Technol. Jan., 83 – 88

Rosengger, L., Wu., R.1990, Integrated PVT data characterization- lessons from four case histories, J. Can. Per. Tech., 38, (13), (paper 97-05)

Shojaei, M., Bahrami, F., Barati, P., Riahi, S., 2014, Adaptive neuro-fuzzy approach for reservoir oil bubble point pressure estimation, *J. of Natural Gas Sci. and Eng.*, *Volume 20, September 2014, Pages 214-220.*

Soave, G. 1972. Equilibrium constants from a modified Redlich-Kwong equation of state, Chem. Eng. Sci. 27, 1197-1203.

Standing, M.B. 1947. A Pressure-Volume-Temperature Correlation for Mixtures of California Oils and Gases. *API Drilling and Production Practice* (1947): 275-287.

Tale, M., Alquraishi, A., Al Mushaigeh, E. 2009, New model for crude oil saturation pressure prediction using alternation conditional expectation, Offshore Mediterranean Conf. & Exhibit. Italy 25-27, 2009.

Talebi, R., Ghiasi, M. Talebi, H., Mohammadyian, M., Zendehboudi,S., Arabloo, M., Bahadori, A., 2014, Application of soft computing approaches for modelling saturation pressure of reservoir oil, *J. Nat. Gas Sci. and Eng., Vol. 20, Sep. 2014, 8-15.*

Vazquez, M., Beggs, H. D. 1980, Correlations for Fluid Physical Property Prediction. *J Pet. Tech.* 32 (6): 968-970. Vazquez, M.E. 1976. *Correlations for Fluid Physical Property Prediction,* MS thesis, University of Tulsa.

Velarde, J., Blasingame, T.A., and McCain Jr., W.D. 1997. Correlation of Black Oil Properties at Pressures below Bubble Point Pressure - A New Approach. Presented at the Annual Technical Meeting of CIM, Calgary, Alberta, 8–11 June. PETSOC 97-93.

Vogel, J.R., Yarborough, L. 1980, the effect of nitrogen on phase behavior and phase properties of reservoir fluids, SPE 8815.

Whitson, C. H., 1980, Characterizing hydrocarbon plus fraction, SPEJ, 24 (4), 683-694.

Williams, C.A., Zana, E.N., 1980. Use of the Peng-Robinson equation of state to predict hydrocarbon phase behavior and miscibility for fluid displacement. SPE 8817, SPE/DOE on Enhanced Oil Recovery, April 20–23, Tulsa, OK.

Appendix

In this appendix, sample calculations are presented to show the deficiency of some of the published methods.

Input parameters:

N_2	CO_2	H_2S	C_1	C_2	C_3	C_4	C_5	C_6	C_{7+}	GR.C7+	MWC7+	T, °F	Ps, psia
0.65	0.02	0	45.02	12.45	8.93	6.03	3.02	1.44	22.44	0.81	184	140	3002

Ahmadi et al (2014)

Ahmadi et al. proposed an 8 steps calculation (copied from their paper) to estimate the BPP as follows:

$$\mathrm{BPP} = A \times Mw_{C7+} - \frac{B}{C_{7+}} + C \times \mathrm{Vol./Inter.} + D \times SG_{C7} + \frac{E}{SG_{C7+}} + \frac{F}{Mw_{C7+}} + T \times 505.967 + \frac{2.45789 \times 10^6}{T} + \frac{G}{\mathrm{Vol./Inter.}} \quad (5)$$

$$A = 1488.46 - SG_{C7+} \times 1087.64 - \frac{601.512}{SG_{C7+}} + 1670.58 + T \times 0.40771 + \frac{6345.73}{T} + \frac{8.72863}{\mathrm{Vol./Inter.}} \quad (6)$$

$$B = \frac{4132.53}{\mathrm{Vol./Inter.}} + \frac{6.1977 \times 10^7}{Mw_{C7+}} + T \times 3181.39 + \frac{1.06393 \times 10^8}{T} \quad (7)$$

$$C = \frac{20964.7}{C_{7+}} - \frac{1.76162 \times 10^6}{Mw_{C7+}} - T \times 19.1963 - Mw_{C7+} \times 21.5427 + \frac{12851.5}{SG_{C7+}} - \frac{712613}{T} - 3477.49 + SG_{C7+} \times 9532.73 \quad (8)$$

$$D = 1.05345 \times 10^{+06} - \frac{1.50748 \times 10^8}{Mw_{C7+}} - T \times 676.048 - \frac{1.20252 \times 10^7}{T} - \frac{12178.2}{\mathrm{Vol./Inter.}} \quad (9)$$

$$E = 641357 - \frac{1.02891 \times 10^8}{Mw_{C7+}} - T \times 301.034 - \frac{1.91446 \times 10^6}{T} - \frac{3873.62}{\mathrm{Vol./Inter.}} \quad (10)$$

$$F = 2.51471 \times 10^{+08} + T \times 25332.5 + \frac{3.47893 \times 10^8}{T} - \frac{244573}{\mathrm{Vol./Inter.}} \quad (11)$$

$$G = 20125.6 - T \times 23.9271 - \frac{703193}{T} \quad (12)$$

Ahmadi et al. define the Vol. /Inter. as the mole percentage ratio of the volatile (C_1& N_2) to the intermediate components (C_2 to C_6, CO_2 and H_2S). Application of the above equations from their paper using the input parameters results in the following:

Vol./Inter.	A	B	C	D	E	F	G	BPP
1.43211	1643.9	1545062	-271.127	45122.71	23642.47	2.57E+08	11753	1,794,123

The calculated saturation pressure (BBP) from Eq.5 is 1,794,123 psia which is unreasonable. This indicates a deficiency in the model.

Gholami et al (2014)

Gholami et al. presented 3 steps calculation (copied from their paper) to estimate the BPP as follows:

$$Tr(M) = \sum_{i=0}^{5} B_i M^i \quad (15)$$

Application of Eq. 15 resulted in the following

N_2	0.00584
CO_2	0.060917
H_2S	0.050636
C_1	-0.47658
C_2	-0.41547
C_3	-0.45083
C_4	-0.02003
C_5	0.068769
C_6	0.273921
C_{7+}	1.814523
SG.C7+	-0.0197
MWC7+	37.03525
T., F	-0.11139

Application of Eq. 16

$$Tr(P_b) = Tr(N_2) + Tr(CO_2) + Tr(H_2S) + Tr(C_1) + Tr(C_2) + Tr(C_3) + Tr(C_4) + Tr(C_5) + Tr(C_6) + Tr(C_{7+}) + Tr(SG_{C7+}) + Tr(MW_{C7+}) + Tr(T) \quad (16)$$

Resulted that Tr(Pb) = 37.8159

$$P_b = (57.0840387741 \times (Tr(P_b))^2) + (1166.9102304199 \times (Tr(P_b))) + 2207.3573204358 \quad (17)$$

Application of Eq. 17 resulted in saturation pressure (P_b) = 127,967 psia, which is unreasonable, compared to the measured one is 3002 psia. This indicates that there is a deficiency in the model.

Jarrahian et al (2015)

Jarrahian et al. proposed the following Eq. (copied from their paper) to estimate the saturation pressure:

$$P_{sat} = A_0 \exp\Big(A_1 z_{N_2} + A_2 z_{CO_2} + A_3 z_{H_2S} + A_4 z_{C_1}^{A_5} + A_6 \sum_2^5 z_{C_i} + A_7 z_{C_{7+}}^{A_8} \gamma_{C_{7+}}^{A_9} Mw_{C_{7+}}^{A_{10}} + A_{11}(T - 459.67)\Big) \quad (2)$$

Application of Eq. 2 resulted in saturation pressure (Ps) = 1812 psia, compared to the measured one is 3002 psia. This indicates that model has inaccuracy problem. This has been the pattern in all the crude oil samples shown in Table 4 and figures 6A through 6D.

Table 1- Summary of attempts to calculate the saturation pressure of crude oils

Method based on production information	Method based on compositional data
Standing, 1947;	Elsharkawy , 2003
Elam, 1957;	Tale et al., 2009
Lasater, 1958;	AlQuraishi, 2009
Vasquez, 1976;	Bandyopadhyay and Sharma, 2011
Vasquez and Beggs, 1980;	Khamehchi et al., 2011
GlasØ, 1980;	Farasat et al., 2013
Labedi, 1982&1992;	Ahmadi et al., 2014
Owolabi, 1984;	Gholami et al. 2014
Al-Marhoun, 1985;	Jarrahian et al 2015
Obomanu and Okpobiri, 1987;	
Asgarpour et al., 1989;	
Al-Najar et al., 1988;	
Dokla and Osman, 1990;	
Petrosky, 1990;	
Kartoatmodjo and Schmidt, 1991&1994;	
Macary and El-Batanoney, 1992;	
Omar and Todd, 1993;	
Hasan, 1993;	
De Ghetto and Villa, 1994;	
Farashad et al., 1996;	
Almehaidaeb, 1997;	
Elsharkawy, 1997;	
Velarde et al., 1997;	
Kairy et al., 1998;	
Petrosky and Farshad, 1988;	
Al-shammasi, 2001;	
Gharbi & Elsharkawy 1999	
Levitan and Murtha, 1999;	
Dindrouk and Christman, 2001	
Moradi et al., 2010	
Okoduwa and Ikiensikimama 2010	
Asoodeh and Bagheripour, 2012	
Shojaei et al., 2014	
Talebi et al., 2014	
Elmabrouk et al., 2014	
Kanu and Ikiensikimama, 2014	
Jarrahian et al 2015	

Table 2- Description of the data used to develop and test the extended model

Input	Training model (179 Data)			Testing (28 Data)			Testing (7 new experimental)		
	Min	Ave	Max	Min	Ave	Max	Min	Ave	Max
N_2	0.00	0.35	1.67	0.00	0.27	1.67	0.10	0.26	0.35
CO_2	0.00	1.27	11.37	0.00	1.17	11.37	0.14	0.41	0.83
H_2S	0.00	0.15	3.68	0.00	0.15	3.22	0.00	0.00	0.00
C_1	13.16	33.86	74.18	19.50	34.79	73.36	20.64	28.25	42.79
C_2	3.36	8.07	13.71	4.64	8.26	11.63	6.69	7.88	10.82
C_3	0.89	6.59	11.87	2.46	6.76	12.03	6.05	6.59	7.10
C_4	0.95	4.46	8.4	1.66	4.41	6.58	4.03	5.05	5.93
C_5	0.40	3.21	5.95	1.00	3.11	5.23	3.36	4.33	5.52
C_6	0.00	3.14	6.37	0.99	3.15	5.46	2.88	3.74	4.33
C_{7+}	10.72	38.89	57.73	11.18	37.88	51.22	27.61	43.49	51.31
$SG.C_{7+}$	0.74	0.87	0.959	0.70	0.86	0.93	0.85	0.89	0.92
MWC_{7+}	134	240	368.9	105	223	292	216	262	298
T, °F	58	156	319	58	165	261	121	157	230
Ps, psia	537	2198	5065	1025	2307	5000	1227	1779	3344

Table 3: Error analysis of the proposed models in comparison with all the investigated methods

Method	ARE	AARE	SD	R^2%
PR-EOS	6.24%	10.19%	1028	97.9
SRK-EOS	-3.12%	10.36%	1024	96.7
Elsharkawy (2003)	0.85%	8.57%	1015	96.4
Alquraishi (2009)	-6.06%	10.16%	1088	96.7
Bandyopadhyay (2011)	8.05%	12.28%	1054	96.8
Farasat (2013)	NA	NA	NA	NA
Gholami et al. (2014)	?	?	?	?
Ahmadi et al. (2014)	?	?	?	?
Jarrahian et al (2015)	35.52	35.56	983	95.2
New Extended Model	-0.37%	8.20%	1030	97.6
New Lumped Model	-0.50%	8.46%	1036	97.4

ARE = Average Relative Error

AARE = Average Absolute Relative Error

SD = Standard Deviation

R^2 = Correlation Coefficient

NA = Calculation was not possible, because the weight and bias were not provided.

? = The method produces unreasonable results as indicated in the appendix.

Table 4- Comparing model accuracy for calculations of saturation pressures for various crude oils

Volatile-Oil-Samples (C7+<30)																						
Fluid Source	index	N2	CO2	H2S	C1	C2	C3	C4	C5	C6	C7+	GR.C7+	MWC7+	T, °F	Ps, psi	SRK	PR	Elsh	Alq	Bandy	Jarr	Extd	Lump
Rosenegger	7	0.03	8.39	0	47.4	10.3	6.12	4.2	2.89	2.05	18.6	0.83	180	295	4000	3933	3879	4460	4420	4165	3317	4139	4178
Vogel-80	13	0.65	0.02	0	45	12.5	8.93	6.03	3.02	1.44	22.4	0.81	184	140	3002	2624	2548	2742	2895	2643	1812	3019	2996
Dansh-92	18	0	0	0	46.8	8.77	7.44	4.01	2.56	4.02	26.4	0.77	158	212	2941	2550	2594	3248	3617	3269	1944	3244	3374
	31	0.44	0.38	0	49.1	7.6	6.13	3.84	2.52	2	28	0.84	231	199	3739	3522	3507	3418	3951	3508	2309	3625	3644
Middle East	45	0.03	0.97	0	41.6	10.8	7.52	4.5	3	2.43	29.1	0.85	208	241	3069	3109	2979	3080	3513	3138	1962	3127	3140
Coats-86	51	0.55	1.03	0	36.5	9.93	8.85	6	3.78	3.56	30.4	0.84	200	234	2746	2537	2472	2714	2945	2643	1682	2777	2720

Black-Oil -Samples (35<C7+<55)																						
Fluid Source	index	N2	CO2	H2S	C1	C2	C3	C4	C5	C6	C7+	GR.C7+	MWC7+	T, °F	Ps, psi	SRK	PR	Elsh	Alq	Bandy	Jarr	Extd	Lump
Coats-86	73	1.64	0.08	0	28.4	7.16	10.5	8.4	3.82	4.05	36	0.84	252	131	1694	1674	1618	1615	1738	1644	1084	1713	1713
Middle East	88	0.08	1.37	0.69	36	8.67	5.94	2.97	1.5	3.1	39.7	0.9	274	168	2505	2710	2318	2247	2608	2321	1476	2441	2445
Middle East	114	0.29	0.48	0	28.4	8.29	7.38	5.06	3.42	4.41	42.3	0.88	252	133	1632	1672	1477	1544	1748	1517	984	1587	1598
Middle East	122	0	1.25	0	33.4	9.24	6.07	2.4	1.62	2.78	43.3	0.85	252	230	2110	2313	2278	2424	2737	2438	1410	2388	2412
Middle East	134	0	0.17	0	28.6	8.63	7.43	3.81	2.61	4.11	44.7	0.86	249	134	1548	1551	1429	1477	1764	1433	937	1503	1567
Middle East	146	0.44	0.83	0	27.8	7.54	6.9	4.61	3.1	3.11	45.7	0.9	272	133	1565	1799	1505	1558	1730	1528	1002	1602	1590
moharam-96	149	0.21	0.34	0	20	7.93	8	6.6	5.87	5.08	45.9	0.86	230	235	900	1353	1290	1531	1514	1225	773	1303	1294
Middle East	162	0.35	0.47	0	26.5	7.71	6.05	4.03	4.08	4.22	46.6	0.89	253	135	1500	1613	1380	1427	1631	1391	923	1478	1456
Middle East	176	0.12	0.51	0	28.8	7.91	6.46	3.17	2.07	3.25	47.7	0.88	250	134	1615	1665	1476	1574	1783	1502	981	1571	1614
Middle East	194	0.21	0.15	0	27.8	7.68	6.19	3.41	1.81	2.84	49.9	0.86	228	134	1590	1532	1377	1545	1680	1426	923	1466	1493
Rosenegger	207	0.37	0.02	0	17.3	6.11	8.23	5.47	3.76	3.96	54.8	0.85	242	156	537	938	888	914	1064	836	558	661	731

Newly-Measured Kuwait University Laboratory																							
Fluid Source	index	N2	CO2	H2S	C1	C2	C3	C4	C5	C6	C7+	GR.C7+	MWC7+	T, °F	Ps, psi	SRK	PR	Elsh	Alq	Bandy	Jarr	Extd	Lump
KU-1	209	0.33	0.22	0	25.6	6.87	6.39	5.61	4.68	4.33	46	0.879	222	133	1595	1360	1530	1442	1514	1394	860	1357	1337
KU-2	210	0.35	0.47	0	26.5	7.71	6.05	4.03	4.08	4.22	46.6	0.893	253	135	1500	1400	1620	1427	1631	1391	924	1478	1456
KU-3	211	0.27	0.74	0	23.1	7.2	6.45	5.93	4.66	4.16	47.5	0.906	290	138	1283	1140	1360	1284	1448	1296	800	1287	1219
KU-4	212	0.22	0.26	0	20.6	6.69	6.65	5.11	5.52	3.6	51.3	0.919	298	171	1227	1120	1300	1188	1415	1170	742	1176	1131
KU-5	213	0.19	0.14	0	27.8	8.37	6.88	5.28	4.33	3.85	43.2	0.89	278	121	1524	1300	1560	1402	1659	1409	928	1534	1489
KU-6	214	0.33	0.24	0	31.3	7.48	6.58	4.86	3.7	3.17	42.3	0.9	274	171	1978	1800	2200	1975	2242	1985	1224	2060	2026
KU-7	215	0.1	0.83	0	42.8	10.8	7.1	4.51	3.36	2.88	27.6	0.85	216	230	3344	3070	3400	3083	3568	3139	2003	3213	3200

Low-Temperature Oil Samples (59°F<T<71°F)																							
Fluid Source	index	N2	CO2	H2S	C1	C2	C3	C4	C5	C6	C7+	GR.C7+	MWC7+	T, °F	Ps, psi	SRK	PR	Elsh	Alq	Bandy	Jarr	Extd	Lump
Middle East	101	0	1.12	0.21	27	10.8	9.23	4.97	2.62	3.33	40.8	0.88	249	71	1247	1090	1033	1040	1235	564	808	1148	1205
Middle East	117	0.36	0.17	0	27.2	8.93	8.6	6.07	3.71	2.55	42.4	0.88	271	68	1130	1028	1245	1094	1248	766	820	1239	1228
Middle East	135	0	0.17	0	28.6	8.63	7.43	3.81	2.61	4.11	44.7	0.86	249	70	1270	1055	1255	1130	1311	783	826	1229	1297
Middle East	138	0.14	0.54	0	29.4	7.96	6.19	3.96	3.25	3.84	44.7	0.9	290	59	1265	1065	1350	1191	1301	880	894	1408	1390
Middle East	147	0.44	0.83	0	27.8	7.54	6.9	4.61	3.1	3.11	45.7	0.9	272	67	1320	1125	1400	1201	1268	921	879	1319	1312
Middle East	173	0.06	0.81	0	31.3	7.42	5.68	2.82	1.63	2.64	47.6	0.89	270	68	1460	1215	1535	1441	1463	1026	975	1548	1576
Middle East	193	0.16	0.3	0	24.7	7.51	6.92	4.76	2.65	3.43	49.6	0.88	239	71	1050	975	1156	1008	1107	701	714	958	986

Elsh: Elsharkawy (2003) Alq: AlQuraishi (2009) Band: Bandyopadhyay and Sharma (2011) Jarr: Jarrahian et. al (2015)
Extd: Extended model Lump: Lumped model

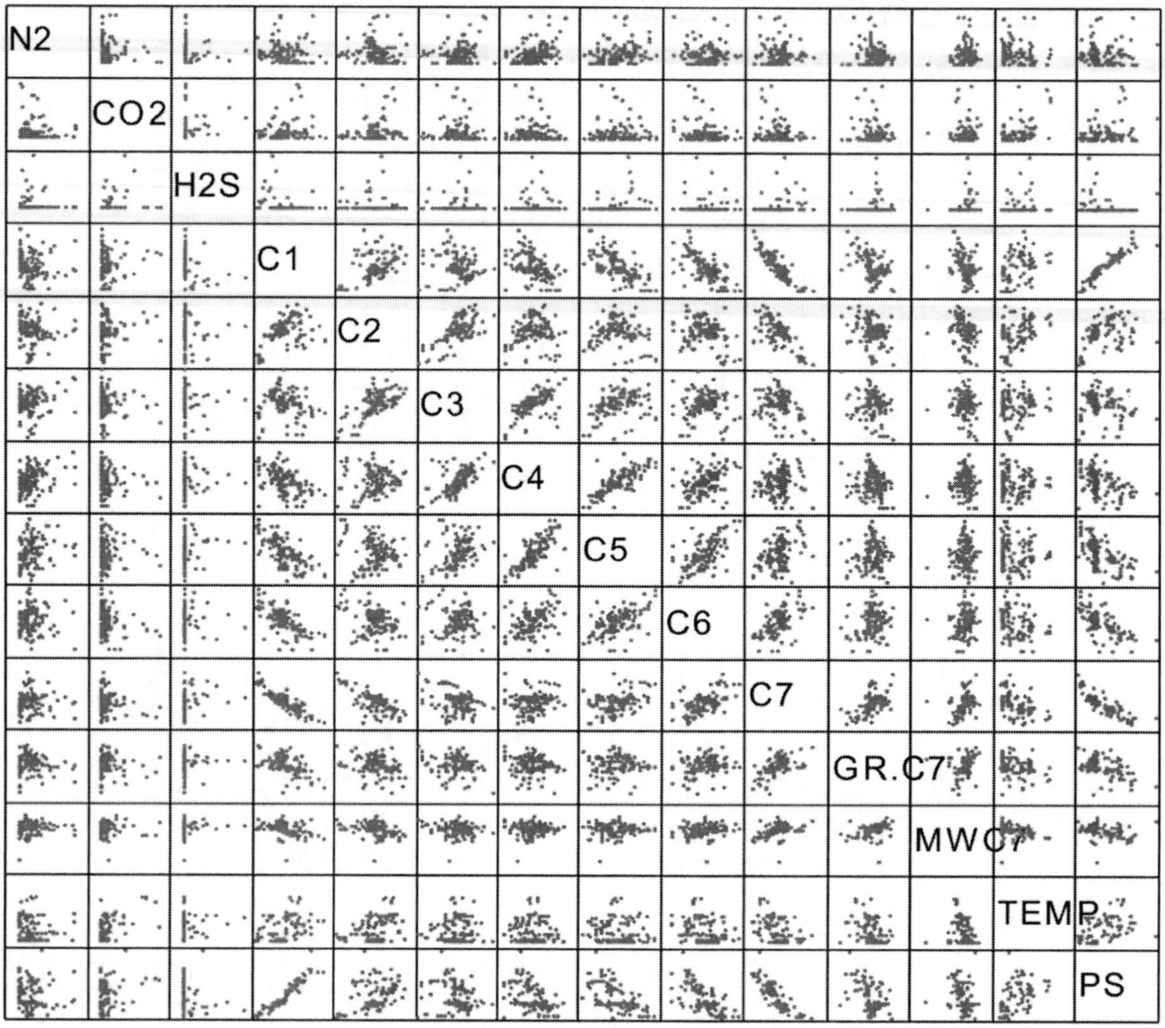

Figure 1: Correlation matrix for all the variables in the oil data bank

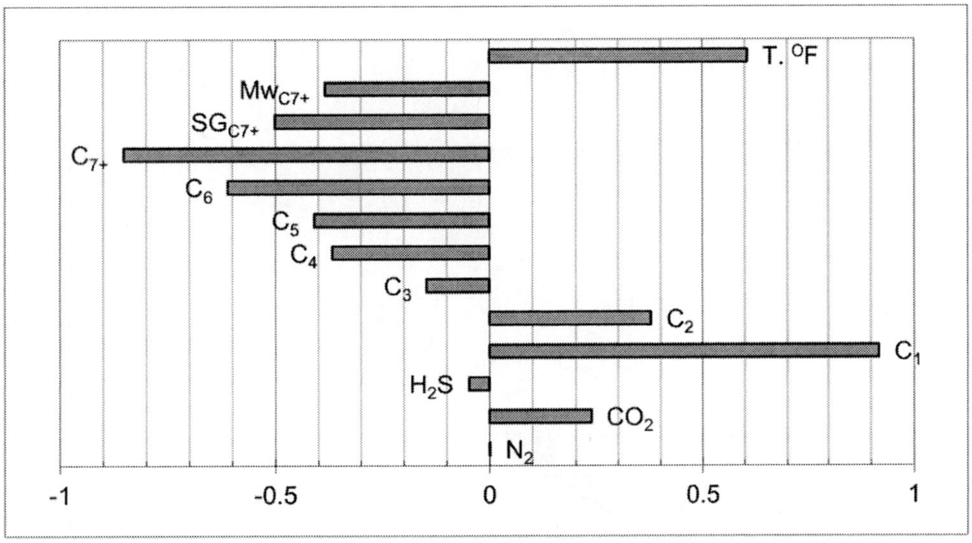

Figure 2: Correlation coefficient for all the variables in the training model (179 data)

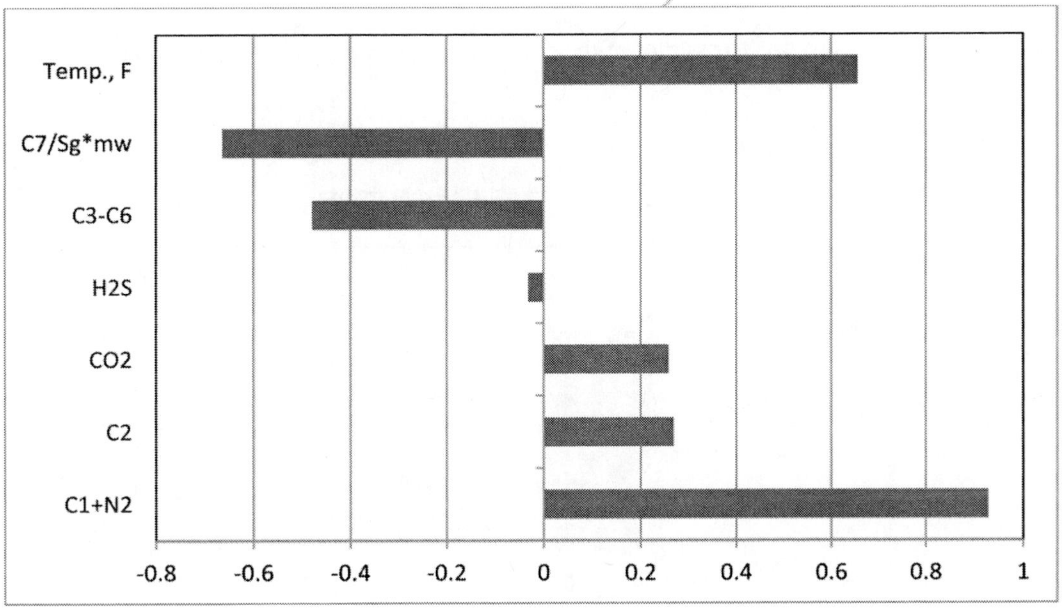

Figure 3: Correlation matrix for the lumped model (179 data)

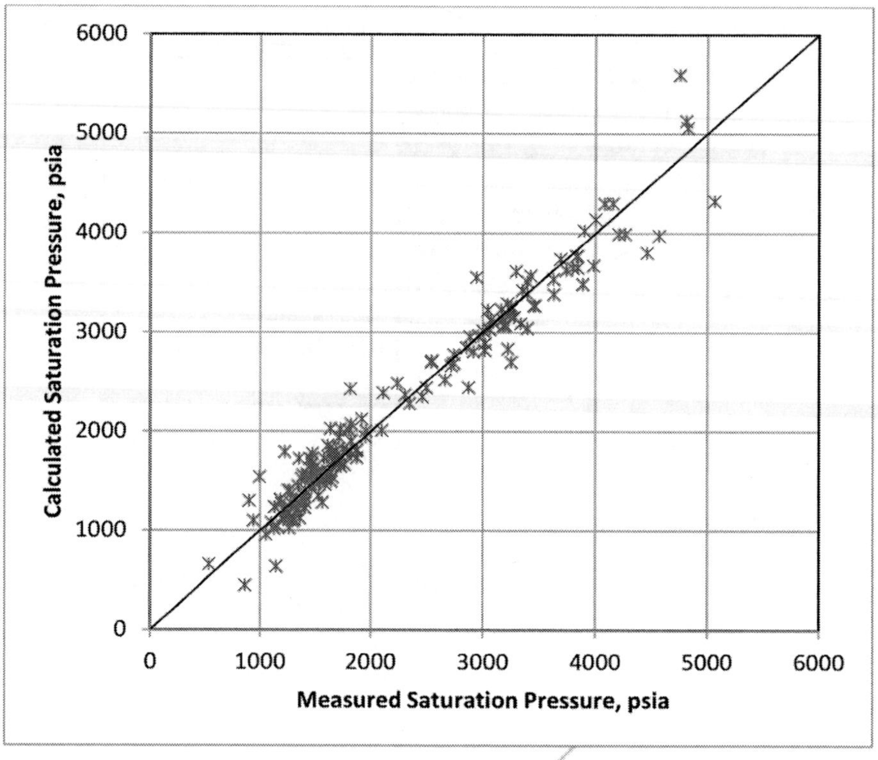

Figure 4A: Cross plot of the extended model using 179 training data

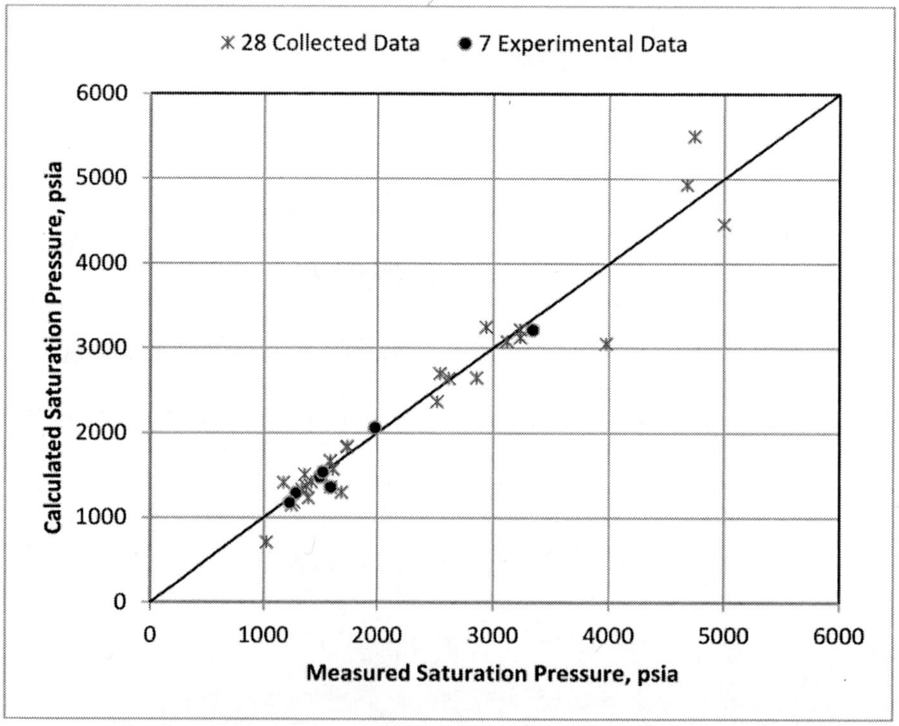

Figure 4B: Cross-plot of the extended model using 35-testing data

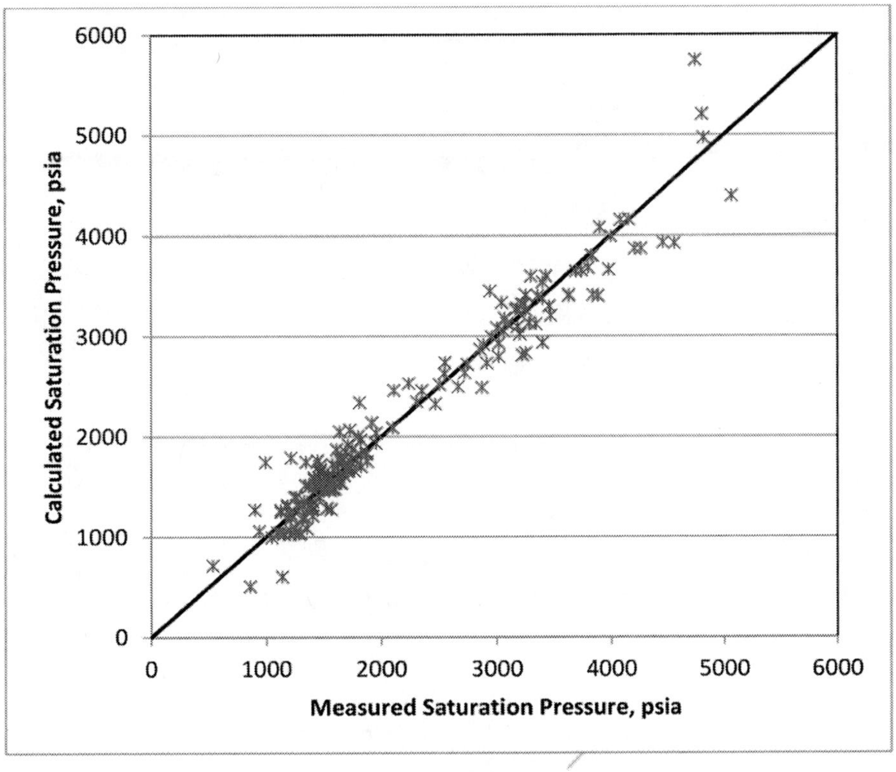

Figure 5A: Cross-plot of the lumped model using 179 training data

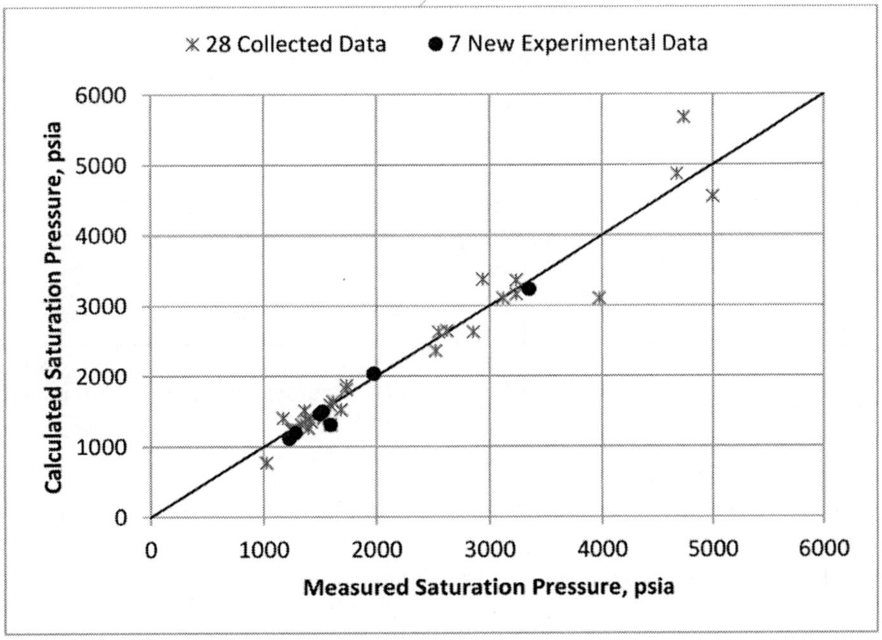

Figure 5B: Cross-plot of testing data for the lumped model

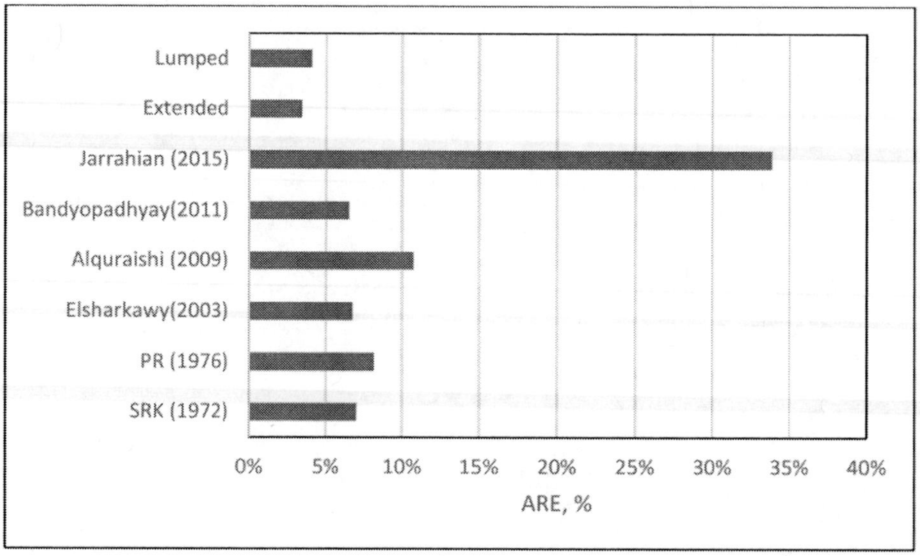

Figure 6A: Comparison of all the models' accuracy for the volatile oil samples

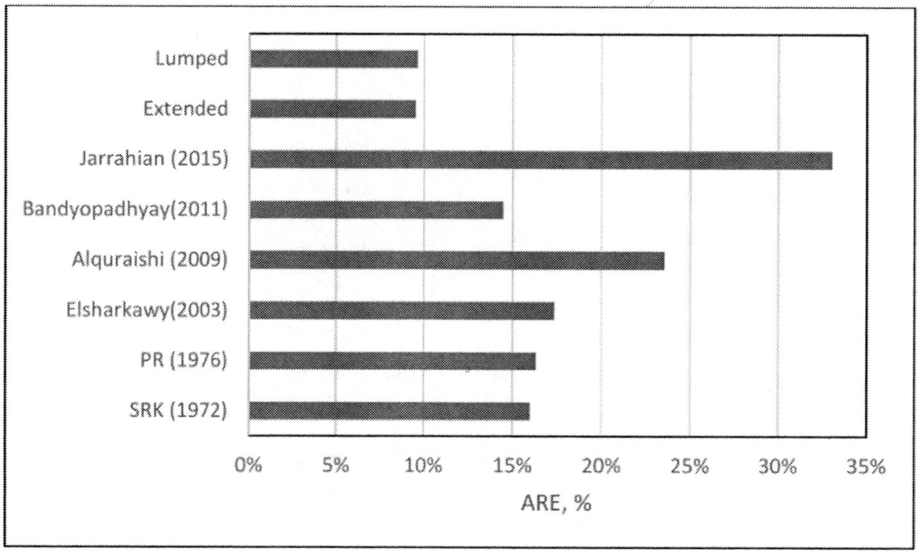

Figure 6B: Comparison of all the models' accuracy for the black oil samples

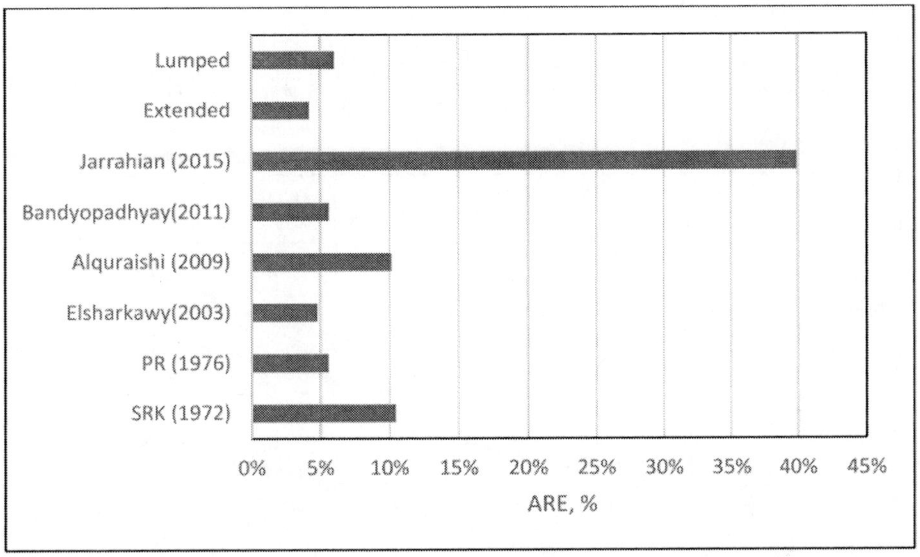

Figure 6C: Comparison of all the models' accuracy for the newly measured samples at KU

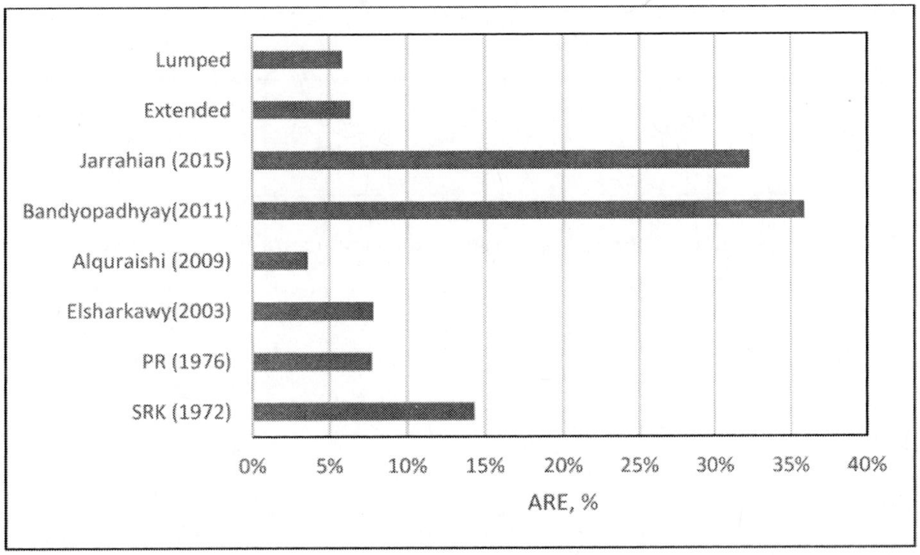

Figure 6D: Comparison of all the models' accuracy for low-temperature oil samples

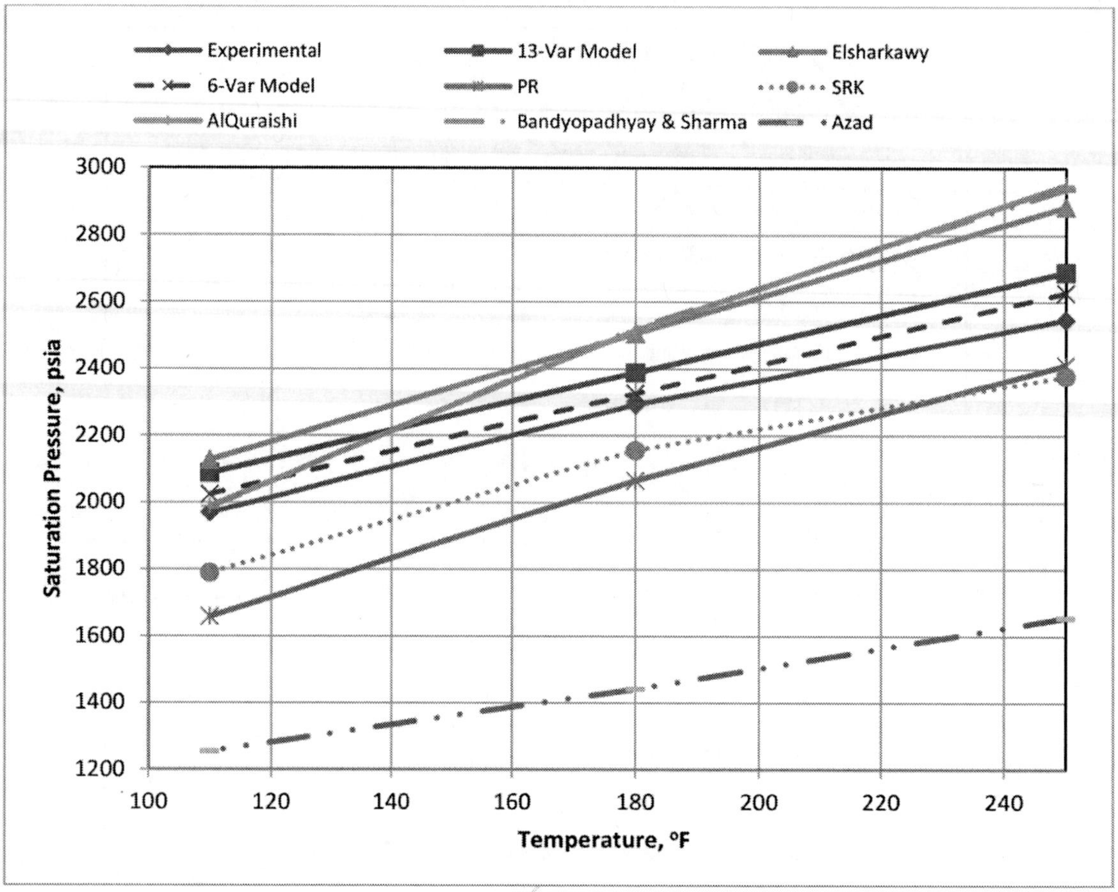

Figure 7: Changes in saturation pressure as a function of temperature for the different models (Sample #70)

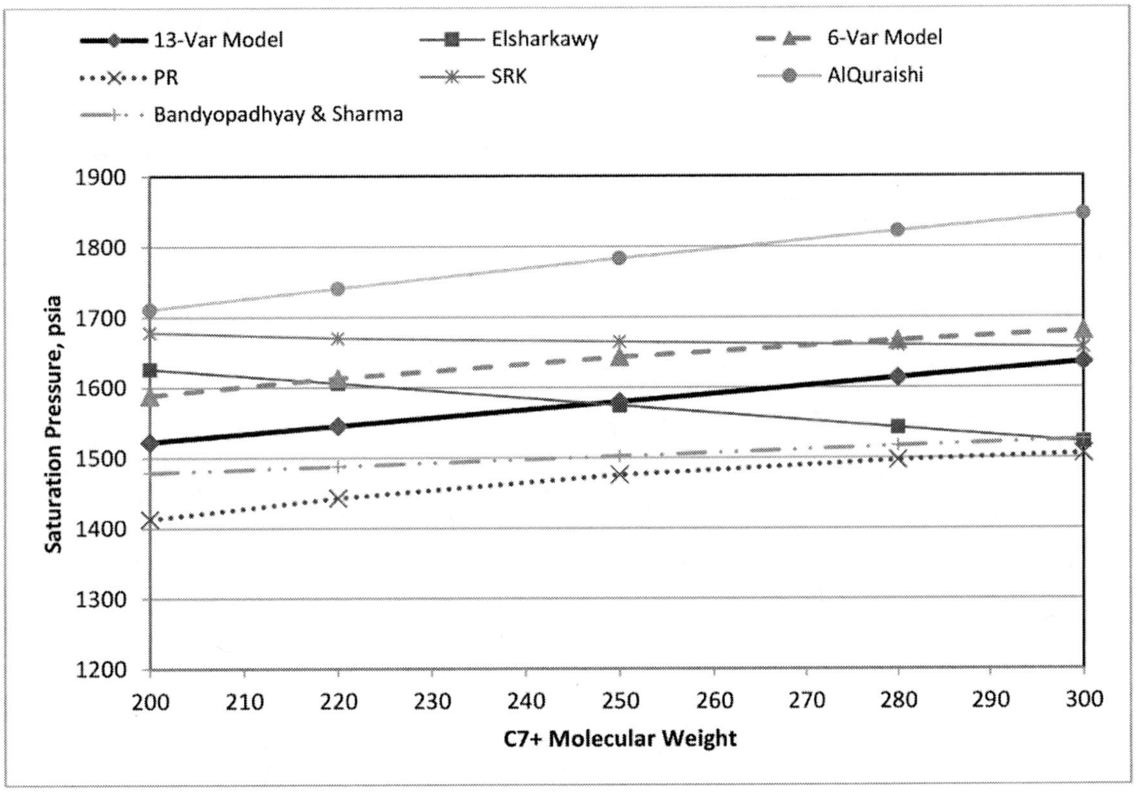

Figure 8: Comparing model performances: Changes in BPP as a function of C_{7+} molecular weight at 134°F (Sample #176)

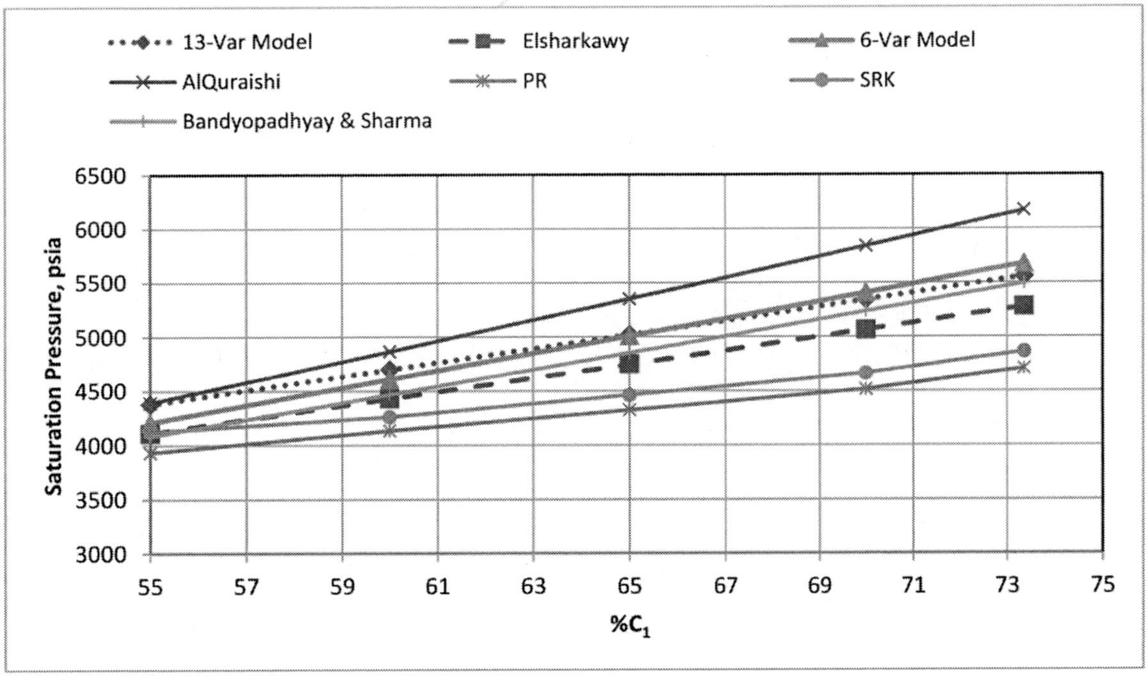

Figure 9: Sensitivity of saturation pressure to methane content at 212°F (sample #2)

444834 Stability and Rheology of Emulsified Acid in Waste Oil: Effect of Emulsifier Concentration, Mixing Speed, Temperature and Droplet Size

Abstract:

Acid in oil emulsion is used in the industry for stimulation purposes. In particular, HCl as the acid phase and diesel as the oil. Emulsifying HCl has numerous benefits over regular HCl. Perhaps the main reason is due to deeper penetration. Apart from that, less corrosion damages are caused since the external phase is hydrocarbon. Several studies showed the success of replacing diesel as an external phase with other hydrocarbon oil, such as crude oil and xylene.

This work utilizes the extra hydrocarbon left unused –or sometimes dumped- from refineries. This unused hydrocarbon is referred to as waste oil. First the chemical composition of waste oil was studied using HPLC, SARA fractionation as well as FTIR. This helped in selecting the proper emulsifier agent for the waste oil. The HCl-waste-oil emulsion was prepared using 15wt% HCl with a ratio of 70:30 of acid-to-oil. Extensive work was done to ensure reproducibility of data. This study was done to experimentally investigate the effects of distinct important variables on thermal stability and apparent viscosity of the HCl-waste-oil emulsion. These variables are; emulsifier concentration, mixing speed and the addition of corrosion inhibitor. All experiments were conducted at a high temperature 120°C. To assess the quality of the HCl-waste-oil emulsion, droplet size was measured as a function of temperature (up to 100 °C).

The results showed that the HCl-waste-oil emulsion is a shear-thinning fluid. Power-law model was applicable seamlessly to all of the apparent viscosity data for all of measured temperatures. The optimum conditions were found to be; 0.5vol% and 600 rpm for emulsifier concentration and mixing speed, respectively. Apart from that, the emulsions prepared with corrosion inhibitor were less stable compared to the ones without it. However, the addition of corrosion inhibitor tended to generate monodispersed emulsion but bigger size droplets for temperatures 25°C and 50°C and polydispersed droplets for higher temperatures. Droplet size measurements showed an average size of 1.47 to 3.09 μm. Overall, the lab results showed a promising potential for the HCl-waste-oil emulsion to be used in the field.

This work takes into account; reducing the budget of acidizing job and at the same time environmental concerns. It makes use of the unwanted hydrocarbon from refiners. And, it saves the environment by using the waste oil instead of dumping it.

443836 Oxidative Dehydrogenation of n-Butane on Carbon Catalysts

Abstract:

Butenes and butadiene, which are one of the important industrial chemical products to produce synthetic rubber , are sythesized traditionally by oxidative dehydrogenation (ODH) of n-butane over various mixed metal oxides. Metal-free carbon nanomaterials has recently received much attention by researchers as a catalyst for the ODH reaction due to its environmental acceptability, corrosion resistance, and unique surface properties. For the reason, the reaction system was modelled in this study for the sythesis of butene and butadiene. For the modelling of the reaction system, the catalytic performance of several carbon catalysts, such as graphite and activated carbon, were initially tested. Then, the reaction mechanisms and kinetics were developed based on experimental results. Lastly, the mathematical modeling of the reaction system was optimized to maximize an overall yield of the main products. This approach allowed us to develop the optimized reaction system more efficiently.

This research was respectfully supported by Engineering Development Research Center (EDRC) funded by the Ministry of Trade, Industry & Energy (MOTIE). (No.N0000990)

443837 Blowdown Dynamic Modeling and Analysis of Pressure and Temperature

Abstract:

Vessel blowdown is one of the safety operation procedures that discharges flammable gas or liquid to flare system during emergency situation such as fire. When blowdown occurs, the liquefaction and freezing effect cause a pressurized vessel to be brittle, exposing it to possible hazard situation. Also the blowdown systems are recommended to be accordance with API(American Petroleum Institute) 521 code. API 521 requires pressure in vessel should be half of the initial pressure or be lower than 100 psig in 15 minutes after commencing blowdown in fire case. Therefore, accurate estimation over the change of temperature and pressure of the vessel during blowdown period is crucial for the design of blowdown system. For the reason, researchers have developed a few dynamic blowdown models so far. However, the model was found to be rather inaccurate under certain conditions. Therefore a new dynamic blowdown model was developed in this work. For the accurate analysis, the model algorithm considers both vapour and liquid phase in non-equilibrium condition that the temperatures of vapour and liquid phases are different. Futhermore, heat transfer from the vessel wall to discharge gas at either laminal or turbulent flow in the vessel was considered in the model to increase the accuracy of blowdown estimation. For thermodynamics, both Peng-Robinson and Soave-Redlich-Kwong equations of state were applied to the model, and the results were compared each other. Finally, the simulation results were compared with Haque et al's experimental and steady-state simulation results.

444364 Molecular Simulation of Controlling Ions and Their Role in Low Salinity IOR for Carbonate Reservoirs

Abstract:

Carbonate geology represents around 50% of hydrocarbon-bearing reservoirs in the world and more than 70% in some areas such as the Middle East. Water flooding has not been very efficient in carbonate reservoirs due mainly to the low water wetness. Modifying the ionic composition and salinity of the injected seawater demonstrated that the oil recovery from these carbonate reservoirs could be improved. Although the underlying mechanisms are still a subject of debate, the improved oil recovery has been mostly attributed to the increased water-wetness of the rock through the interactions of specific ions like Ca^{2+}, Mg^{2+}, SO_4^{2-}, and Cl^- , which are present in seawater with the rock surface under favorable conditions

The study of hydrated gas-phase metal ions provides a link between intrinsic chemistry of the isolated ion and its chemistry in solution. The structures and binding enthalpies of a variety of gas-phase and solution mono- and divalent ion hydrates containing up to 12 water molecules have been studied computationally. A DFT and molecular dynamics simulations were employed to study hydrated structures of the form $[\mathbf{M}(H_2O)_n]^{2+}.mH_2O$ for calcium and magnesium and $[\mathbf{M}(H_2O)_n]^{+}.mH_2O$ for sodium, in which there are *n* water molecules in the first coordination shell and *m* water molecules in the second coordination shell. The hydration, dehydration, and interactions of metal ions with Cl^{-1} and SO^{-2}_4 have been studied to better understand the mechanism of low salinity IOR. Density functional (DFT) theory using the B3LYP functional, and 6-311++g** basis set has been used. The charge of the metal ion decreases monotonically as the number of water molecules increases in the complex. Extensive calculations of the energetic contribution of individual structural characteristics as well as the energetic trends in binding energy are developed, which will lead to better understanding of how low salinity water IOR works.

Keywords: Hydration, carbonate reservoir, DFT, binding energy, Low Salinity

447119 Development R&D of Manoir Industries and Emphasis on Coating

Abstract:

The Senior VP Technical Support & Projects will present the development R&D of Manoir Industries and especially the coating (it is our lattest development) as well the services we can provide such as metallurgy expertise, advice, convector cleaning, site welding, welding supervision in order to increase the furnace productivity. So it is 60% R&D and 40% services. Manoir Industries manufactures cast components for the petrochemical market and develops high temperature alloys such as the Manaurites.

442925 Identifying Gas Liquid Contactor Performances for Optimized Natural Gas Desacidification Amines Units - from Process Calculations to Experimental Strategy

Abstract:

Amine units are widely used for gas sweetening. The absorber performances of these reactive absorption processes are typically mass transfer limited. The proper selection of gas liquid contactor is thus critical to optimize the size of towers, the solvent flowrate and the duty of the reboiler. This work proposes an approach to identify the most suitable packings for a given application.

The case-study selected here to illustrate the approach deals specifically with a CO_2removal application where the sour gas is at 7 % vol, and the sweet gas specification is 2% vol. The gas flowrate is fixed at 0,18 MMSCFD with a solvent flowrate fixed at the minimum of the thermodynamic constraints (via equilibrium curves). An in-house rate-based simulator with accessible forms of mass transfer correlations, based on a double film model, is used to calculate the absorber height and diameter which are needed to reach the sweet gas specifications. A reference structured packing is compared with standard valve trays.

The calculations show that the bed height required with the first contactor is considerably higher than with the second one while the column diameter is reduced. In this case the use of standard valve trays is more attractive for cost optimization. Interestingly, because mass transfer is accelerated, the absorption rate with the structured packing is found to be thermodynamically limited due to a more severe temperature increase which reduces the driving force in the bottom section of the absorber.

The study determined that the optimum design for the investigated case is a compromise between a full trayed tower, that generally imposes a larger diameter, and a packed tower where the higher mass transfer forces result in a temperature rise in the bottom of the tower which in turn imposes a height increase and a much higher solvent flowrate. . To determine the mass transfer characteristics of a gas-liquid contactor providing this optimum, a sensitivity analysis has been conducted. Mass transfer parameters, i.e., liquid and gas side mass transfer coefficients and interfacial area, were varied by a factor 2 relative to the ones obtained with the reference structured packing. Calculations show that one should use a contactor that reduces liquid phase resistance, specifically, the mass transfer conductance $k_L a_e$ should be increased by at least 50%. This target could be reduced to +30% if one could simultaneously reduce the gas phase conductance $k_G a_e$ by 20%. This result can be explained by the link between mass and heat transfer (Chilton-Colburn analogy) whereby reducing $k_G a_e$decreases the temperature level which in turn thermodynamically increases the mass transfer driving force.

The next step will be to review the literature more extensively to make a short list of pre-selected random and structured packings that best meet the targets. Experimental tests will then be run on IFPEN facilities to measure flooding limits and mass transfer parameters in a small 150 mm diameter column. This step is required to confirm packing's behavior, considering the disparity of the values reported by different authors in the literature. The last step is to consolidate the existing correlations for the most promising packing, and mitigate differences between various authors by performing characterization tests on larger diameter test columns. The final correlations will be implemented in our in-house rate based simulator for industrial use.

The present approach can be applied to determine the required mass transfer characteristics for optimizing other gas treatment applications. However, the conclusions reached here as regards to mass transfer targets and the problematic temperature levels, are valid only to the present case –study, and are not directly applicable to other gas sweetening applications (selectivity, LNG, ...). For example, in a case where H_2S and CO_2 are treated simultaneously the mass transfer limitations could be very different.

443784 Measurement and Correlation of the Mutual Solubility of Diisopropylamine + Water and Triethylamine + Water Systems at High Pressure

Abstract:

The mutual solubility data of diisopropylamine + water and triethylamine + water systems were measured at the pressure up to 100 MPa and the temperature range was 280.2~316.2 K. Experiments were performed in a high-pressure autoclave with a moveable piston. The experimental data were correlated satisfactorily with NRTL activity coefficient model and the binary parameters were obtained.

444161 Novel Microwave Synthesis Carbon Dots from PINE Bark and Its Application for Heavy Metal Sensing

Monday, April 11, 2016
Exhibit Hall E (George R. Brown)

Saheed Sanni, Chemistry, Vaal University of Technology, Vanderbijlpark, South Africa, Puthalapattu ReddyPrasad, Chemistry, Vaal University of Technology, Vanderbiljpark, South Africa and Augustine E. Ofomaja, Chemistry, Vaal university of Technology, Vanderbiljpark, WV, South Africa

NOVEL MICROWAVE SYNTHESIS CARBON DOTS FROM PINE BARK AND ITS APPLICATION FOR HEAVY METAL SENSING

Sanni S.O., ReddyPrasad P., Ofomaja A.E.

Biosorption and Waste water Treatment Research Laboratory, *Department of Chemistry, Faculty of Applied and Computer Sciences, Vaal University of Technology, P. Bag X021, Vanderbijlpark-1900, South Africa.*

INTRODUCTION

Microwave pyrolysis for the production of carbon has the advantages of being cheap, produces high yield of carbon and is absolutely a green synthetic method [1]. In recent times, the use of microwave pyrolysis have been extended to the preparation of carbon dots. Carbon nanodots (C-dots) falls under the class of carbon nanoparticles with size below 10 nm[2]. C-dots are carbon materials which processes high water solubility and strong fluorescence depending on their size, edge, shape, surface ligands, and defects. Since C-dots preparation can be achieved from difference carbonaceous sources and by varying a range of parameters during their synthesis, the properties of the final product may vary depending on these factors. C-dots have sp^2 characteristics and possess several functional groups on their surfaces that makes them highly hydrophilic and easily functionalized with various organic, polymeric, inorganic, or biological spices[3]. Several synthesis methods that have been explored includes electrochemical synthesis, combustion/thermal/hydrothermal/acidic oxidation, supported synthesis, microwave/ultrasonic, arc discharge, laser ablation/passivation, and plasma treatment[4]. C-dots processes excellent such as water dispersibility, chemical stability and photostability, ease of modification, excitation-dependent multicolor emission, low toxicity, and good cell permeability, great efforts have been paid to their potential applications in biosensing, bioimaging, drug delivery, and other biological related aspects.

METHODS

The carbon dots (C-dots) with green luminescence was synthesized from microwave assisted pyrolysis of pine bark waste our carbon source at 800, 1000 1200 W for 1 Hr. The dots produced were characterized using FTIR, TGA, XRD, UV-Vis and Raman spectroscopy. The dots were applied for heavy metal sensing.

RESULTS AND DISCUSSION

The functional groups present on the C-dots (**Fig.1**) includes -OH stretching band mode at 3413 cm^{-1}, C-H stretching at 2934 and 2845 cm^{-1}, stretching peaks of C=O, C=C at 1731, 1604 and 1437 cm^{-1}. The XRD pattern (**Fig 2a &b**) of produced C-dots depicts sharp peak at 20.9° and 26.8°, which is highly crystalline cores and possess amorphous nature of C-dots. This peculiar peak is ascribed to 002 of the graphitic carbon, which in line with inter-planar spacing of 3.77 Å with higher spacing than bulk graphite at 3.34 Å. The thermogravimetric and differential thermal analysis display (**Fig 3**) that the C-dots possess thermal stability more than 100 °C and mass losses occurs afterwards. The thermal analysis depicts 3 endothermic peaks at 331, 542 and 735 °C, which are ascribed to oxygenated groups such as carboxyl, hydroxyl, carbonyl and phenols. The optical properties of the synthesize C-dots were explored using UV-vis absorption and photoluminescence (PL) spectroscopy. The UV-vis spectra display (**Fig 4**)intense absorption peak in the range of 260 to 290 nm, which is ascribed to π-π^* transition of the aromatic sp2 domain and also presence of multiple polyaromatic chromophores. A shoulder peak at 330 nm which is ascribed to n-π^* transitions of C=O and structural defects of the C-dots. PL spectrum of the C-dots (**Fig 5a**)exhibits an intense maximum emission peak at 432 nm upon excitation at 330 nm, which also shows characteristic yellowish green color under sunlight and intense blue coloration UV lamp (365 nm). The PL properties of as prepared C-dots was varied at (**Fig5b**) different excitation wavelength from 300 to 405 nm. There was red-shifting in emission peak as the excitation peak was increased from 300 to 405 nm and also simultaneously decrease in fluorescent intensity, which is ascribed to quantum effect and surface defects. This behavior of the prepared C-dots is dependent on excitation-fluorescence, which is accordance with other report [13]. The quantum yield of synthesized C-dots was measured to be 17% at maximum excitation at 330 nm

Fig.6 shows the Raman spectrum of the PB-GQDs, which has been usually used to confirm the quality of the prepared PB-GQDs. There is two major Raman peaks at 1370 cm^{-1} (D band), 1580 cm-1 (G band). The peak of G band corresponds to the Sp^2 bonded carbon atoms in a two-dimensional (2D) hexagonal lattice and the peak of G band is associated with the vibrations of carbon atoms with dangling bonds in the termination plane of tangled graphite or glassy carbon. The D/G ratio band ratio is ID/IG: 0.68 is confirm the formation of graphene like structure from synthesized pine bark.

Fluorescence detection of Cu^{2+} ion

The detection of Cu^{2+} was performed at room temperature in phosphate buffer (PBS) solution pH 4.3. 1.0 mL of PB-CQDs solution was added into 0.5 mL of PBS buffer, followed by the addition of 1.0 mL different concentrations of Cu^{2+}. The fluorescence emission spectra were recorded after reaction for 5 min at room temperature (~28 °C).

Selectivity and sensitivity studies

Moreover sensitivity and selectivity is additional significant parameter to assess the performance of the sensing system. Consequently, we examined the fluorescence intensity changes in the presence of representative metal ions such as including Cu^{2+}, Hg^{2+}, Ca^{2+}, Al^{3+}, Ba^{2+}, Mn^{2+}, Ni^{2+}, Pb^{2+}, Co^{2+}, Zn^{2+}, Fe^{3+}, $Cs^{2+}$and Mg^{2+}, of each at a concentration of 15.0 μg mL^{-1} were reacted with a PB-CQDs solution. Fig. 7 A displays the relative change in PL intensity of PB-CQDs in the presence of various metal ions. Fe^{2+}, Mn^{2+} and Hg^{2+} ions cause the slight PL changes (defined as the relative change of PL intensity in 60–80% as compared to blank) which can be attributed to the nonspecific interactions between the carboxylic and/or amine groups and the metal ions. Among these metal ions, Cu^{2+} displays the strongest PL quenching effect on PB-CQDs, assigning to the special coordination between the Cu^{2+} ions and the phenolic hydroxyl and/or amine groups of PB-CQDs which has been widely used for the detection of Cu^{2+} ions or colored reactions in traditional organic chemistry. Such a specific fluorescence quenching effect may originate from the strong interactions between Cu^{2+} ions and the surface groups of PB-CQDs which transfer the photoelectrons from PB-CQDs to Cu^{2+} ions. These results clearly demonstrate that the PB-CQDs-based Cu^{2+} sensor is highly selective to Cu^{2+} over the other metal ions. Fig.8. Fluorescence responses of CCDs to the addition of different metal ions under acetate buffer (pH 4.3). The concentration of each metal ion is 15 μg $mL^{-1.}$ F_0 and F correspond to the fluorescence intensities of PB-CQDs at emission wavelength of 440 nm excited at 320 nm in the absence and presence of metal ions, respectively.

As shown in Fig. 3, the fluorescence intensity of PB-CQDs gradually decreased with the Cu^{2+} concentration. The fluorescence intensity displayed liner responses with the Cu^{2+} concentrations, with two linear ranges of 0.0□14.0 μg mL^{-1} and 15.0□25.0 μg mL^{-1} (Fig.3B). The linear equations are $F/F0$ =0.9827□0.8485C_{Cu2+} (R = 0.993) and $F/F0$ = 0.7565–0.0033 C_{u2+} (R = 0.996), respectively.

REFERENCES

[1] W. B. Lu, X. Y. Qin, S. Liu, G. H. Chang, Y. W. Zhang, Y. L. Luo, A. M. Asiri, A. O. Al-Youbi, X. P. Sun, Economical, green synthesis of fluorescent carbon nanoparticles and their use as probes for sensitive and selective detection of mercury(II) ions, Anal. Chem. **2012**, 84, 5351-5357.

[2] J. Jiang, Y. He, S. Li, H. Cui, Amino acids as the source for producing carbon nanodots: microwave assisted one-step synthesis, intrinsic photoluminescence property and intense chemiluminescence enhancement. Chem Commun **2012**, 48, 9634–6.

[3] Y. Wang, A. Hu, Carbon quantum dots: synthesis, properties and applications, J. Mater. Chem. C **2014,** 2, 6921–6939.

[4] C. Liu, P. Zhang, X. Zhai, F. Tian, W. Li, J. Yang, et al., Nano-carrier for gene delivery and bioimaging based on carbon dots with PEI-passivation enhanced fluorescence, Biomater. **2012,** 33, 3604-13.

Figures:

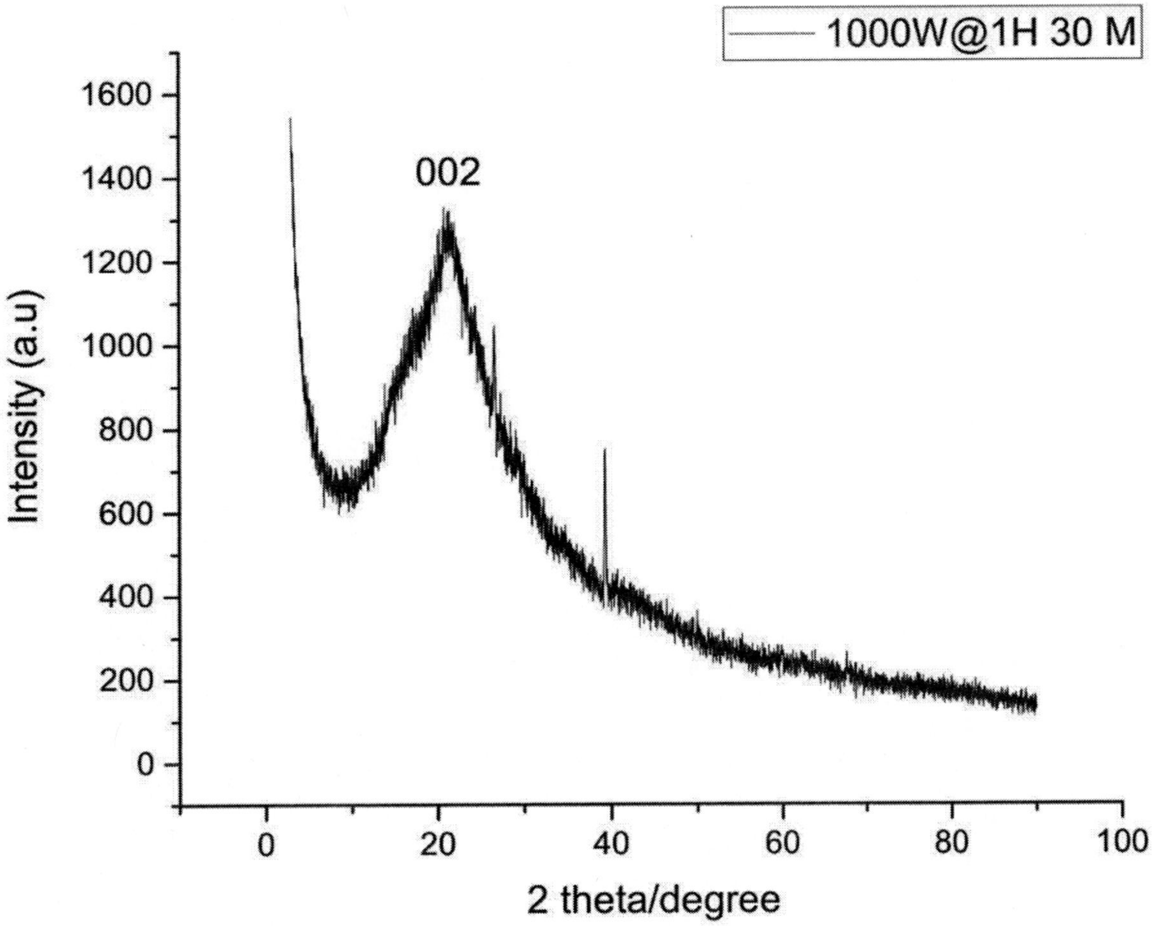
1000W@1H 30 M
002
Intensity (a.u)
1600
1400
1200
1000
800
600
400
200
0
0
20
40
60
80
100
2 theta/degree

Fig 1. FTIR spectra of as-prepared C-dots from Pine Bark

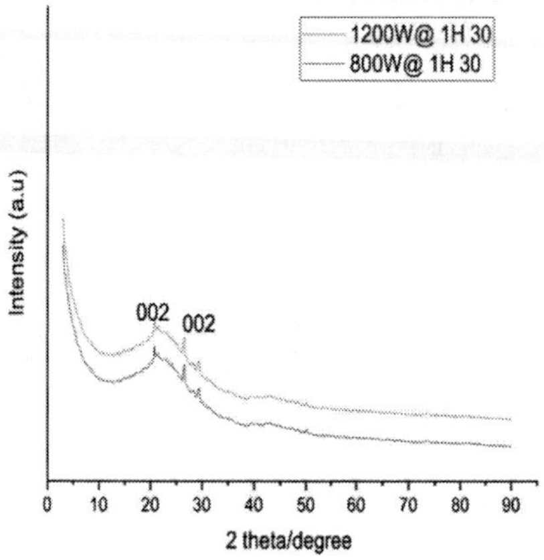

(a)

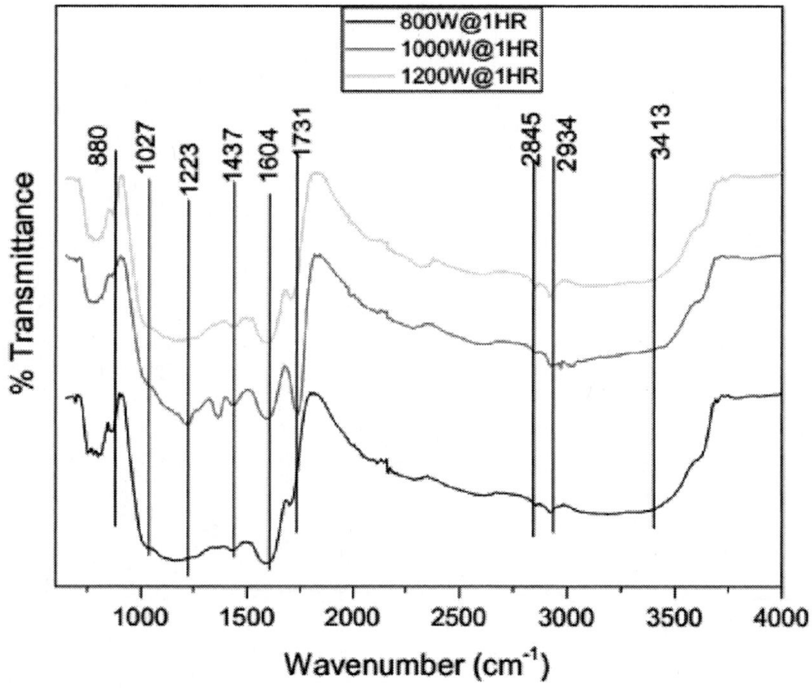

(b)

Fig. 2a & b. XRD pattern of C-dots.

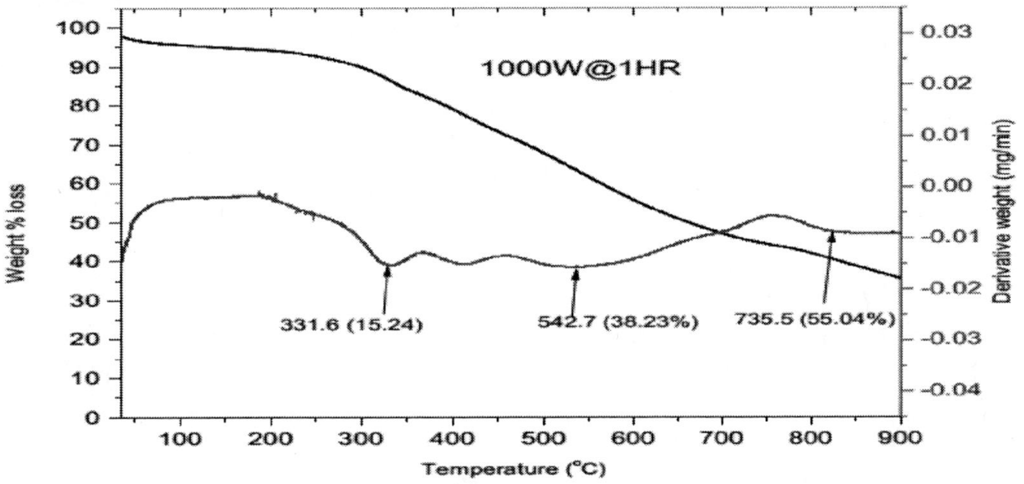

Fig 3. Thermal curves for C-dots under N_2 atmosphere.

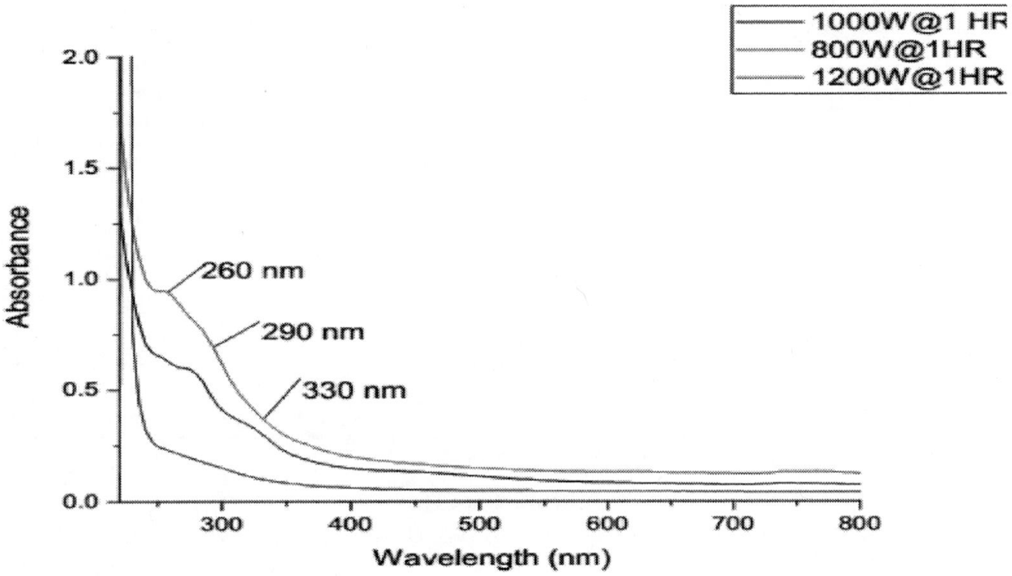

Fig 4. UV-vis absorption of C-dots.

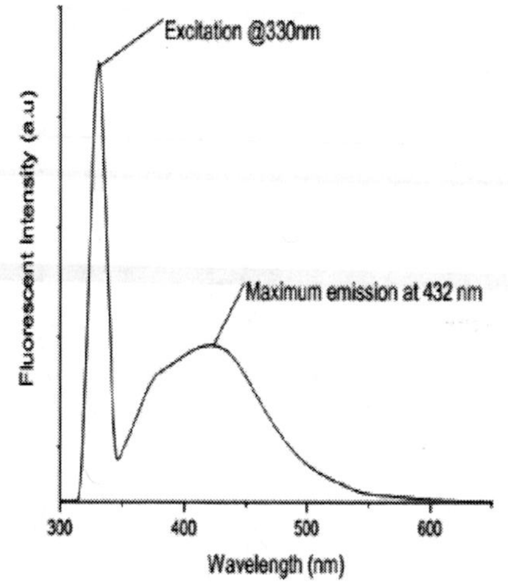

(a)

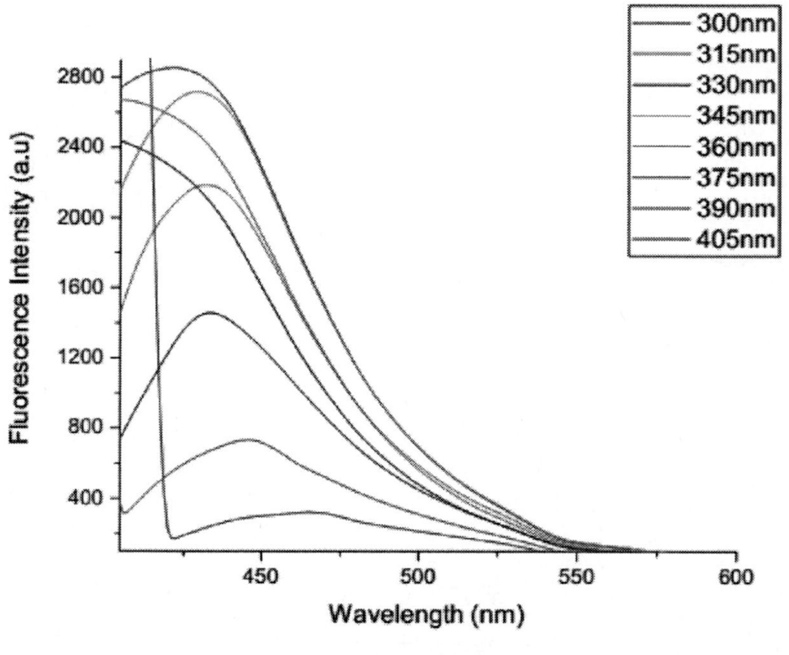

(b)

Fig 5a. PL excitation and emission wavelength (b) PL emission spectra at different excitation wavelength from 300 to 405 nm in 15 nm increments.

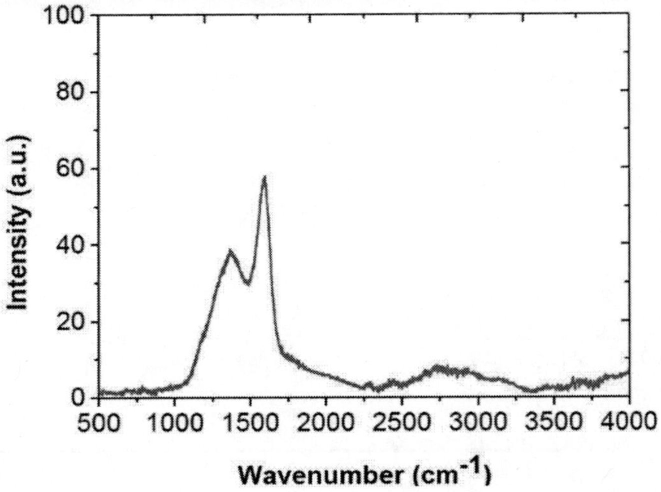

Fig. 6. Raman Spectra of PB-CQDs

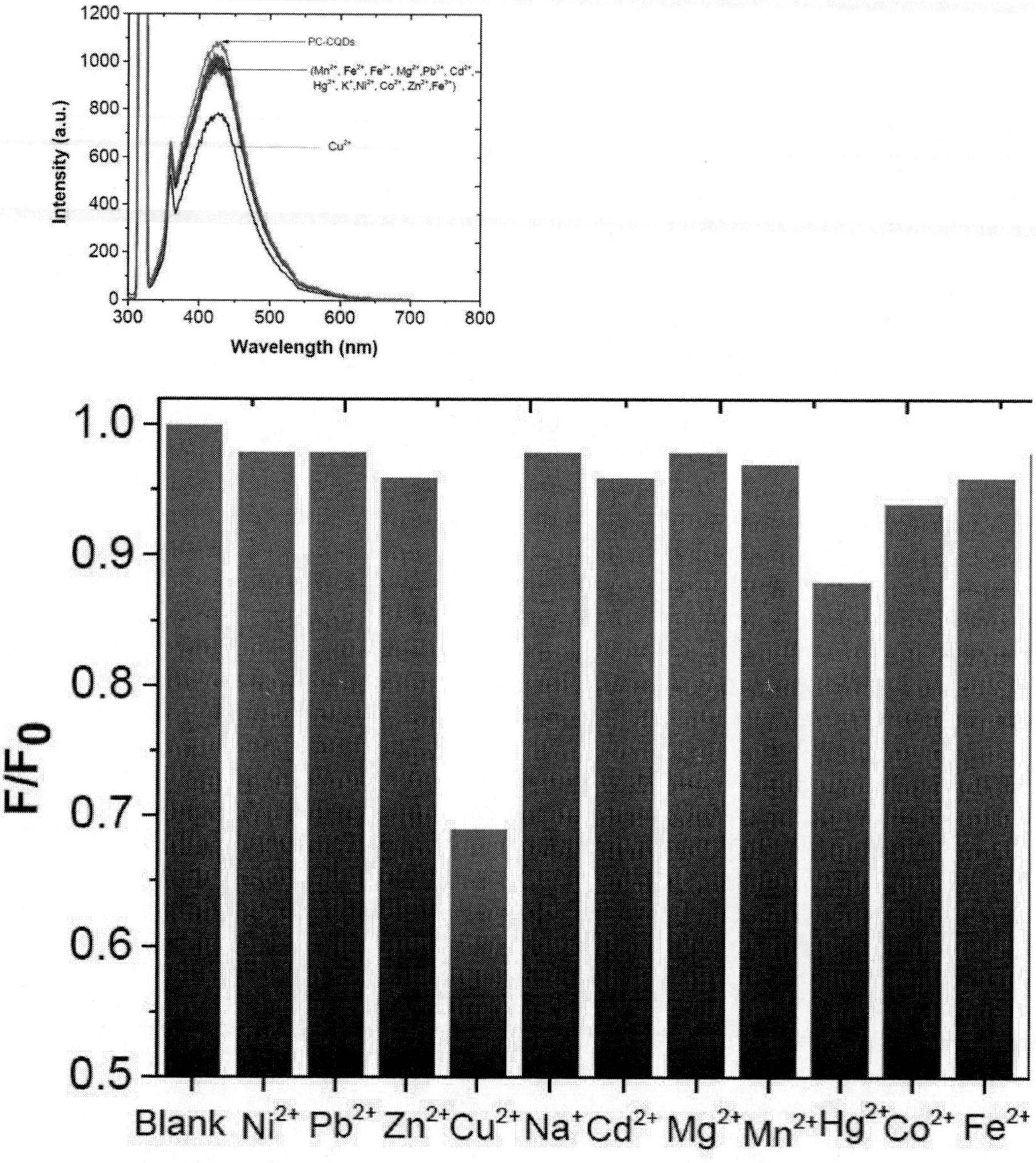

Fig.7. Fluorescence responses of CCDs to the addition of different metal ions under acetate buffer (pH 4.3).

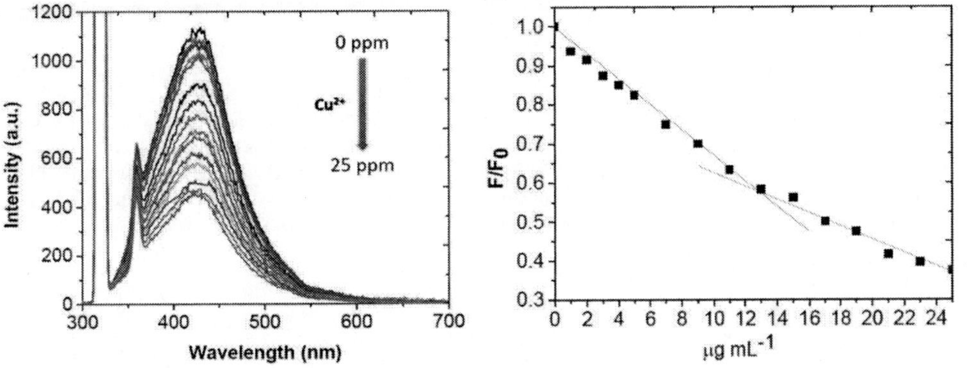

Figure 8. Fluorescence titrations of the y-CDs with Cu^{2+} from 0 to 25 μg mL^{-1}. (Influence of Cu^{2+} concentration on the fluorescence intensity of PB-CQDs (top to bottom: 0, 1, 2, 3, 4, 5, 7, 9, 11, 13, 15, 17, 19, 21, 23 and 25.0 μg mL^{-1})

443095 Experimental and Numerical Study of Transport Phenomenon of Falling Film Flow with Different Heating Methods

Abstract:

Falling film flow is a common phenomenon in chemical industry. It is widely used in chemical instruments, such as structured packed columns and falling film evaporators. An experiment was implemented to study the mass transfer phenomenon of binary-mixture (ethanol-water) falling film flows with different heating methods (uniform or concentrated heating). In order to study the detailed flow and mass transfer behaviors, a 2-D multiphase computational fluid dynamics (CFD) model based on volume of fluid (VOF) was developed. This model divides mass transfer into two parts: one is caused by the vapor-liquid equilibrium (VLE) and the other is caused by the heat flux. Results from simulation are inconsistent with the experimental data. Both experiments and simulations show that the ethanol concentration of the vapor with concentrated heating is 1.5% higher than the one with uniform heating.

443774 Thermodynamic Anaysis of Dehydration of n-Propanol Using Heterogeneous Azeotropic Distillation

Abstract:

Heterogeneous azeotropic distillation is a widely popular process in chemical process industries for the separation of binary azeotropic mixtures into their constituent pure components. It involves the addition of an entrainer which is partially miscible and forms a heterogeneous azeotrope preferably with one of the original components. This is a powerful technique which combines several physical phenomena such as presence of azeotropes and liquid-liquid immiscibilities, and is used widely in industries to enhance difficult or intractable distillation separations. For the design of heterogeneous azeotropic distillation columns, a detailed study of the heterogeneous region is crucial to plan the distillation sequence, which requires experimental vapor-liquid-liquid equilibrium (VLLE) data.

The present study focuses on the investigation of the heterogeneous region of the well known industrially available azeotropic mixture n-Propanol-Water system using Propyl acetate as entrainer. The VLE data of the system was generated by means of modified Othmer still and the still is modified using a probe type ultrasonic homogenizer to account for the heterogeneity. Isobaric vapor- liquid- equilibria and vapor - liquid - liquid - equilibria data at atmospheric pressure have been obtained for the ternary system n-Propanol-water-propyl acetate and the constituent binary systems. The equilibrium data has been correlated using NRTL, UNIQUAC and UNIFAC models and compared with the experimental data.

444269 Functionalization of OH-Terminated Nanoparticles By Cyclic Azasilanes

Abstract:

Colloidal suspensions of functionalized nanoparticles have been suggested for enhanced oil recovery, formation damage remediation, flow assurance, and water treatment operations. Surface modified nanoparticles can also be used as tracers in reservoir studies, active media in thermal management fluids, and fillers in high performance elastomers. Conventional functionalization of nanoparticles with alkoxysilanes and chlorosilanes is limited by needs for catalytic amount of water for the reaction to proceed, requirement of multiple hours to fully saturate the surface, and release of by-products upon reaction completion. Monolayer deposition with a high density of functional groups represents a particularly difficult challenge.

Our ongoing studies indicate that application of cyclic azasilanes that have a strained ring structure with adjacent Si and N centers provides highly efficient functionalization of hydroxylated silica, alumina, magnetite, and diamond nanoparticles which is catalyst-free and by-product-free. The data on synthesis and characterization of functionalized nanoparticles as well as possible reaction mechanism of functionalization will be discussed.

443776 Study of Ionic Liquids As Entrainers for the Extractive Distillation of *N*-Propanol and *N*-Propyl Acetate Binary Azeotropic Mixture and the Isobaric VLE Studies of the System

Abstract:

Distillation is a widely accepted technique for the separation of binary liquid mixtures. Separation by simple distillation is not possible for binary mixtures that form azeotropes. One of the commonly used methods to separate azeotropes is extractive distillation. It is the process of adding an entrainer that can alter the relative volatility of the mixture and thereby breaking the azeotropic point. Ionic liquids on account of their physiochemical properties as well as the recyclability have been used as a preferred replacement for the conventional entrainers. Extractive distillation with ionic liquids has the advantages of both liquid solvent (easy operation) and solid salt(high separation ability). The corrosion and dissolution problems by using inorganic salts can be rectified by these non-volatile, chemically stable, high electrical conductive molten salts.

The azeotropic system under consideration in the present study is *n*-propanol - *n*-propyl acetate mixture. These compounds which constitute the products/reactants of esterification reaction of n-propanol form a minimum boiling azeotrope at 94.7°C. *n*-propyl acetate has been included in a list of volatile acetates as an ecologically benign CFC-free solvent for aerosol sprays.

Vapour liquid equilibrium(VLE) measurements of *n*-propanol and *n*-propyl acetate are performed at atmospheric pressure in a modified Othmer still which is a continuous distillation still in which vapour phase is recirculated. The most suitable ionic liquids for shifting of azeotropic composition of the system are selected using COSMO-RS model with suitable combinations of anions and cations. The isobaric VLE of the system with [EMIM][HSO_4] is obtained and compared with the existing data. The influence of alkyl chain length on the performance of ionic liquids is also analysed with regard to the previous studies. The effect of polarity is the most important factor on the behaviour of ionic liquid solvation in alcohols. Experimental VLE data were correlated using modified UNIFAC model.

444280 Copper-Phyllosilicate Core-Shell Nanocatalysts with Balanced Active Sites for Carbon-Oxygen Hydrogenolysis Reactions

Abstract:

Copper-Phyllosilicate Core-shell Nanocatalysts with Balanced Active Sites for Carbon-Oxygen Hydrogenolysis Reactions

Hairong Yue, Weimao Fang, Changmin Liu, Jiankang Ying

College of Chemical Engineering, Sichuan University, Chengdu, China

Email: hryue@scu.edu.cn

Abstract

Hydrogenolysis of carbon-oxygen (C¨CO) bonds (e.g., esters, ethers, furfural, and CO_2) has emerged as a versatile synthetic tool in organic methodology, as it could produce a variety of products (e.g., chemicals, fuels, and polymers). Copper-based catalysts have been intensively explored for hydrogenation reactions as the copper sites account for the selective hydrogenation of carbon-oxygen bonds and relatively inactive for the hydrogenolysis of carbon-carbon bonds. Work on understanding the active sites of copper catalysts for hydrogenation reactions indicted that both Cu^0 and Cu^+ species were crucial to the activity of Cu-based catalysts. However, researchers were not able to establish a relationship between activity and Cu^0/Cu^+ active species since the conventional Cu-based catalysts is the deactivation by metal particle growth and unstable surface Cu^0 and Cu^+ active species in the strong reductive H_2 and oxidizing carbon-oxygen atmosphere.

We renctly present on efficient approaches for fabrication of a series of copper-phyllosilicate core-shell nanocatalysts and nanoreactors with balanced and stable Cu^0 and Cu^+ active species. We chose the hydrogenation of dimethyl oxalate, ethylene cyclicarbonate and as probe reactions for a better understanding of properties of active sites and the structure-activity relationship. The results indicated a synergy between copper and the oxide components and the balanced surface Cu^0 and Cu^+ species can greatly improve the catalytic C-O hydrogenolysis performance of the catalysts. These nanocatalysts with balanced and stable Cu^0 and Cu^+ active species, confinement effect, intrinsic high surface of Cu^0 and Cu^+ and unique tunable tubular morphology, has potential applications in the high-temperature hydrogenation reactions.

Keywords: core-shell nanocatalysts, Hydrogenation reaction; copper catalyst

References

1. Yue, H.; Zhao, Y.; Ma, X.; Gong, J.: Ethylene glycol: properties, synthesis, and applications. *Chem Soc Rev* 2012, 41, 4218-4244.

2. Gong, J.; Yue, H.; Zhao, Y.; Zhao, S.; Zhao, L.; Lv, J.; Wang, S.; Ma, X.: Synthesis of Ethanol via Syngas on Cu/SiO2 Catalysts with Balanced Cu^0-Cu^+ Sites. *J Am Chem Soc* 2012, 134, 13922-13925.

3. Zhao, S.; Yue, H.; Zhao, Y.; Wang, B.; Geng, Y.; Lv, J.; Wang, S.; Gong, J.; Ma, X.: Chemoselective synthesis of ethanol via hydrogenation of dimethyl oxalate on Cu/SiO_2: Enhanced stability with boron dopant. *J Catal* 2013, 297, 142-150.

4. Yue, H.; Ma, X.; Gong, J.: An Alternative Synthetic Approach for Efficient Catalytic Conversion of Syngas to Ethanol. *Acc Chem Res* 2014, 47(5):1483-1492

5. Yue, H.; Zhao, Y.; Zhao, L.; Lv, J.; Wang, S.; Gong, J.; Ma, X.: Hydrogenation of dimethyl oxalate to ethylene glycol on a Cu/SiO2/cordierite monolithic catalyst: Enhanced internal mass transfer and stability. *AIChE J* 2012, 58, 2798-2809.

6. Yue, H.; Zhao, Y.; Zhao, S.; Wang, B.; Ma, X.; Gong, J.: A copper-phyllosilicate core-sheath nanoreactor for carbon¨Coxygen hydrogenolysis reactions. *Nat Commun* 2013, 4, 2339.

443819 Improving High Shear Rate Friction Reduction Behavior of Polyacrylamide-Based Fracturing Fluids Using Xanthan Gum

Abstract:

The main component of slickwater in fracturing low permeability formations is a high molar mass polymer such as polyacrylamide (PAM), which acts as a drag reduction agent in turbulent flow. Polyacrylamide does not degrade in the environment and also suffers from mechanical degradation at high turbulent flow rates. Five different concentrations of PAM as well as mixtures of polyacrylamide/xanthan were prepared to investigate the possibility of improving shear stability of PAM. Drag reduction measurements were performed in a flow loop and friction factors at different flow rates were calculated using rheological parameters. Based on the observations, Xanthan (XG) is not as good a friction reducer as PAM, even at three-fold higher concentration. XG slightly improved the shear stability in mixed polymer solutions at low concentrations. But at high concentrations, XG significantly improves the friction reduction behavior of PAM at nonionic and ionic environments.

Stimuli-responsive Rubber in Oil and Gas Applications

Rostyslav Dolog, Sankaran Murugesan, and Valery Khabashesku

Baker Hughes, 14990 Yorktown Plaza Dr, Houston, Texas, 77040, USA

ABSTRACT

A problem that exists today in the oil & gas industry is the loss of mechanical properties by reactive rubber seals due to swelling, and the limited space in downhole well applications to accommodate larger swellable packing elements.

This paper presents cost-effective, reinforcing, stimuli-responsive fillers, such as certain classes of fly ash, for use in rubber composite materials in various oil and gas applications. Apart from cost effectiveness, these fillers can also enable stimuli-responsive effects, as well as improved mechanical properties (e.g. modulus) for rubber composites when swelled in water. Fly ash (Class C) is a common component of Portland cement which was used in the literature to achieve a stimuli-responsive effect when added to Hydrogenated Nitrile Butadiene Rubber (HNBR) as filler. Exposure to water brines can cause a filler to promote an increase in pH which enables the hydrolysis of HNBR nitrile groups to carboxyl groups, thus creating an ionomer. This process is followed by neutralization of the new ionomer with metal ions, and formation of ionic crosslinks, further improving mechanical properties of the rubber.

This paper demonstrates that incorporation of fly ash as a reinforcing filler allows development of swellable seals that can retain, or even increase, their modulus and stiffness upon swelling. These characteristics allow avoiding the loss of mechanical properties by seals due to swelling and also eliminating the need for large swellable packing elements that occupy additional space downhole.

KEYWORDS: Elastomers, stimuli-responsive, fly ash, composites

INTRODUCTION

There is a growing demand for high performance elastomers in the oil & gas industry. Rubber compounds are used for various sealing applications, packer elements, stators for pumps, etc. In general, rubber formulations are comprised of elastomer matrix, curing system, antioxidants, plasticizers, along with reinforcing filler, which provides the compounds with their strength. The reinforcing filler is usually represented by carbon black and, to a lesser extent, silica. Along with other factors, filler mechanical properties and concentration generally define hardness, modulus and tensile strength of the elastomer compound. Throughout the years, many different materials have been tried as alternative fillers for rubber composites, including nanomaterials (CNTs, graphene, nanofibers), recycled rubber particles, and ash/fly ash. In the case of ordinary carbon black, the filler serves not only for reinforcement, but also as a cost reducer, because raw elastomer is often a more expensive component. Mechanical properties of the composites are affected by the aspect ratio of the filler, quality of the dispersion, and interface between blend components. The latter determines the effectiveness of load transfer from polymer matrix to the reinforcing filler. High performance compounds can be designed by incorporating well dispersed fine particles with a high aspect ratio and good compatibility with the polymer matrix.

In this context, the fly ash is a cheaper, low reinforcing filler alternative to carbon black, which will help reduce cost of the compounds for "non-demanding" sealing applications requiring large amounts of raw materials. Fly ash is produced in large quantities as a byproduct of coal burned at power plants, and can be considered as a recycled material. Fly ash costs several cents per pound, while carbon black, depending on the grade, is usually one order of magnitude more expensive. Hemmings et al[1] demonstrated that fly ash can be used as filler for polymeric composites with enhanced mechanical properties. Adoption of this technology in the oil and gas industry could change the design approach for reactive rubber compounds. Further functionalization of fly ash through a cost-efficient process could ensure compatibility with the rubber matrix, while promoting filler dispersion and ensuring an efficient load transfer. Such functionalization's include, but are not limited to, surface modification with fatty acids, dicarboxylic acids or phosphoric acid. The reasoning for surface modification is to attach functional groups with aliphatic chains to fly ash particles to make them compatible with aliphatic rubber chains, or to attach acidic groups on their surface that can interact with nitrile groups of HNBR. Initial attempts towards chemical functionalization of fly ash have been made by Seal et al for treatment of offshore oil spills.[2]

Apart from cost effectiveness, composites with fly ash can show stimuli-responsive properties. It has been shown that adding Portland cement to the HNBR matrix improves its mechanical properties in the swollen state by water uptake.[3] Exposure to water causes Portland cement to grow in volume and mass, and promotes an increase in pH which enables the hydrolysis of HNBR nitrile groups to carboxyl groups, thus creating an ionomer. This process is followed by neutralization of the new ionomer with metal ions and formation of ionic crosslinks, further improving mechanical properties of the rubber. Fly ash (Class C) is a common component of Portland cement, which contains a large variety of metal oxides, and is capable of increasing the

---

[1] Filler Comprising Fly Ash For Use In Polymer Composites. Hemmings et.al., US 6,916,863 B2, July 2005.

[2] Hydrophobic High Surface Area Zeolites Derived from Fly Ash for Oil Spill Remediation, Seal et.al., Environ. Sci. Technol., 2013, 47 (11), pp 5843–5850

[3] Stimuli-Responsive Cement-Reinforced Rubber, Musso et. al., ACS Appl Mater Interfaces. 2014 May 14;6(9):6962-8.

pH of the water solution by hydrolysis, which has a potential to achieve a stimuli-responsive effect when added to HNBR as filler.

Degradation of tensile properties is a common problem for swellable packers and other sealing elements. By selectively using certain classes of fly ash as effective reinforcing filler, one can develop swellable seals that can retain or even increase their modulus and stiffness upon swelling. Therefore, there is a potential to minimize or even avoid the loss of mechanical properties by water swellable reactive packing elements due to swelling. This provides an opportunity to decrease the size of these elements, which are often represented by rather long sections, and hence occupy a significant amount of space downhole. In this paper we have studied the reinforcing effects of fly ash fillers with HNBR compounds as well as stimuli responsive properties of these compounds.

EXPERIMENTAL SECTION

Rubber compounds T1, T2, T3, and T4 are elastomers based on a custom-built peroxide curing system. Formulation F2 was prepared by adding 15 wt. % of fly ash class C to compound F1, which is a commercial swellable NBR sulfur-cured compound.

T series compounds were prepared using a C.W. Brabender® internal mixer operated at 60 RPM for 20 minutes at ~160°F and then molded into sheets and cured at 340°F for 30 minutes in a hot press. Compound F2 was prepared by mixing fly ash with compound F1 in the Brabender mixer for 6 minutes at 60 RPM at ~100°F. Compound F1 is a commercial water swellable formulation that was used as received. F series compounds were cured at 300°F for 60 minutes in a hot press.

Tensile properties of rubber samples were tested using dog bone specimens, according to the ISO37 standard, at room temperature and at 350°F using an MTS® testing system.

Swelling tests in 3% KCl brine and LVT 200 oil were performed to assess the effect of fillers on the chemical resistance of elastomers. Standard one-inch rubber buttons were tested at 200°F at ambient pressure for 6 days. Thermal properties were studied using TGA Q50 and DSC Q2000 instruments.

To analyze the stimuli-responsive effect introduced by fly ash fillers, rubber buttons were aged for 10 days in 3% KCl in the oven at 200°F and then removed from the solution, dried for 30 minutes in the oven and then for 3 hours at room temperature. Samples were later tested in the MTS testing system in compression mode.

Viscoelastic properties of stimuli-responsive samples, based on water swellable compounds, were accessed using the RSA G2 dynamic mechanical analysis (DMA) instrument. This experiment was conducted in a temperature ramp mode of 2°C/minute at 1Hz frequency and 0.2% amplitude strain. A Nicolet 6700 Fourier Transform Infrared (FTIR) spectrometer was used to confirm neutralization of carboxylic groups by metal cations.

RESULTS AND DISCUSSION

The reinforcing properties of fly ash were evaluated at room temperature (Table 1) and at 350°F (Table 2) and compared with N550 carbon black. Sample T3, reinforced with fly ash, demonstrated inferior tensile properties and hardness compared to a compound filled with carbon black. However, at room temperature, fly ash was capable of delivering a reinforcing effect to the rubber compound without sacrificing its elasticity, e.g. the ultimate elongation of elastomer compound remained virtually unchanged after incorporation of 50 parts per hundred rubber (PHR) of fly ash, while its modulus and tensile strength have increased. Therefore, potentially, one can substitute carbon black with high loadings of fly ash in certain applications, thus further decreasing the cost of the rubber product. The summary of mechanical properties is provided in Table 1, E_x denotes modulus of a compound at certain strain (x), i.e. stress at strain.

Sample	N550 CB (PHR)	Fly ash Class F (PHR)	Tensile strength (psi)	E_{25} (psi)	E_{50} (psi)	E_{100} (psi)	Elongation (%)	Hardness Shore A
T1	0	0	829	71	97	121	472	48
T2	50	0	3402	165	273	744	290	68
T3	0	50	1176	94	127	166	467	54

Table 1. Room temperature tensile properties and hardness of elastomer samples

The relatively low increase in modulus and hardness of T3 compared to T1 at room temperature can be explained by a low aspect ratio of fly ash and weak interactions between blend components. The latter problem can be solved by functionalization of fly ash particles.

The general data trends from the high temperature tensile tests closely resemble the results of the tensile test conducted at room temperature. However, at high temperature, rubber specimens break at much lower elongation, which leads to a great decrease in the measured tensile strength. This behavior is typical for elastomers and limits their application to a certain maximum service temperature and is also due to the fact that high temperature tests are conducted at conditions that push the limits of the material. The modulus values at 15% (E_{15}) and 20% (E_{20}) strain are not provided herein, because half of the tested specimens were broken before they reached such strains.

The superiority of carbon black as an efficient filler became even more pronounced at high temperature. However, incorporation of fly ash doubled the tensile strength and increased the elongation at break by one-third when compared to the unfilled elastomer.

Sample	E_5 (psi)	E_{10} (psi)	E_{15} (psi)	E_{20} (psi)	Tensile strength (psi)	Elongation (%)
T1	31	46			91	23
T2	61	130	194	261	832	39
T3	42	73	98	119	173	30

Table 2. Summary of high temperature (350˚F) tensile data for elastomer samples

Thermal properties of samples T1-T3 were assessed with Differential Scanning Calorimetry (DSC) and Thermogravimetric analysis (TGA). TGA allows evaluating thermal stability of samples and to verify the actual composition of a sample from the ash content. The results of the DSC and TGA analysis are summarized in Table 3. During TGA, samples were heated to 600°C at 10°C/minute rate in a nitrogen (N_2) environment, and subsequently heated at 10°C/min to 600 °C in air to burn off carbon-based fillers. This allows decomposing of organic matter first by applying heat, and then burn the remaining carbon-based material in air atmosphere.

The initial degradation of the polymer started at 200°C and progressed to complete decomposition of aliphatic chains in the temperature range of 420-500°C with a highest rate of decomposition at ~460°C. Sample T2 incurred the second rapid mass loss after 600°C due to burning of the carbon black, which comprised most of the ash content at 600°C in a N_2 environment. The DSC experiment demonstrated the glass transition temperature of the elastomer remained largely independent of the type of filler - the respective DSC heating scan was conducted with a 10°C/minute temperature ramp after precooling the sample to -50°C.

Sample	W_{600} (%)	W_{800} (%)	$T_{deg}1$, °C	$T_{deg}2$, °C	T_g, °C
T1	2.4	2.4	452.8	-	-29
T2	29.8	2.7	458.7	616.5	-29
T3	32.8	32.7	447.3	-	-28

Table 3. TGA/ DSC results for elastomer samples

Table 3 contains ash contents (W) at 600°C and 800°C, degradation temperatures (T_{deg}) which represent the temperature of the highest rate of decomposition during TGA experiment, and glass transition temperature (T_g) of the samples, as determined by DSC scan.

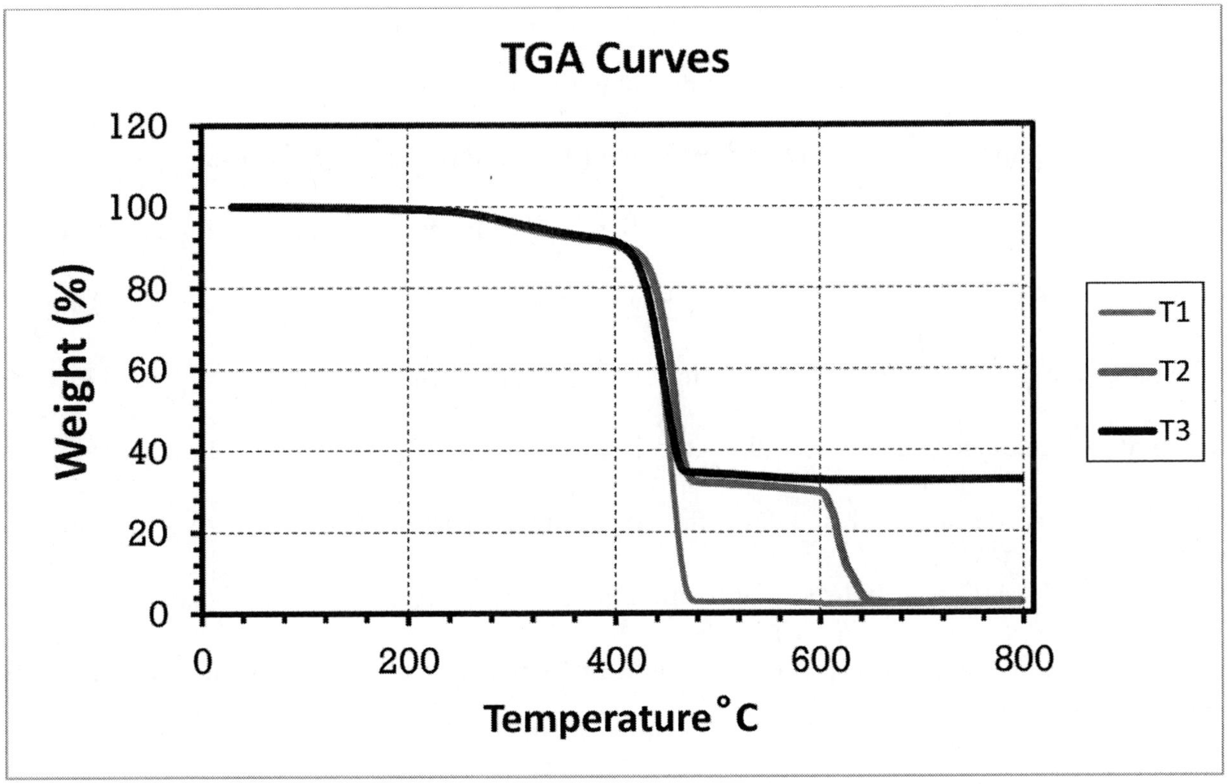

Figure 1. TGA Curves for samples T1, T2, and T3

Sample \ Testing fluid	3% KCl brine	LVT 200	Specific gravity
T1	0.6	13.0	0.96
T2	0.3	11.6	1.13
T3	0.8	9.1	1.18

Table 4. Volume increase (%) of elastomer samples due to swelling for 6 days at 200°F

Musso et. al. developed a stimuli-responsive HNBR composite reinforced with Portland cement that was increasing its stiffness and modulus when swelled in water[3]. Fly ash (Class C) is a common component of Portland cement, which contains the components necessary to achieve a similar effect. Therefore, one can avoid the loss of mechanical properties by water swellable packing element seals upon activation. To test this hypothesis, we studied the effect of aging in water brines on the properties of elastomers with fly ash fillers.

Rubber samples were molded and cured into standard size rubber buttons (cylinders of ~0.25" height and ~1" diameter). These buttons were aged for 10 days in 3% KCl brine in the oven at 200°F, removed from the solution, dried for 30 minutes in the oven and for 3 hours at room temperature. The swelling of both compounds was close to ~1% by volume (Table 4). Samples were later tested in an MTS machine in compression mode. Each button was compressed to 70% strain, buttons remained intact during the experiment, and visually all appeared to have a certain level of compression set at the end of the experiment. The results of the experiments are summarized in Table 5 and Figure 2. Only one button was tested for each sample. The provided compression modulus values (C_n) represent values of stress at a percentage strain (n), e.g. C_5 is stress at 5% compression strain.

Compound T4 was reinforced by fly ash Class C, which contains larger weight fraction of CaO than fly ash Class F (~25% vs ~10-15% respectively) and is expected to produce a higher pH and closely resemble cement that was used to prepare stimuli-responsive rubber in the original paper by Musso et.al.[3] The results demonstrate that exposure to water (3% KCl) at high temperatures can improve the modulus for the compounds filled with fly ash Class C (T4), while the control sample (T1) showed the opposite behavior. After aging, the compressive modulus at 50% strain (e.g. stress at 50% strain) of control sample T1 decreased by 19%, as compared to the unaged sample, while compressive strength of sample T4, reinforced by fly ash, increased by 15%. The difference in modulus values is even more drastic at 70% compression strain (-35% and 31%, respectively). This is in agreement with the hypothesis stated earlier.

Sample	Filler	Compressive modulus (psi)				
		C_5	C_{10}	C_{25}	C_{50}	C_{70}
T1	None	19	42	127	468	1810
T1 swollen	None	16	36	109	380	1185
T4	Fly Ash class C (50 PHR)	21	46	134	434	1628
T4 swollen	Fly Ash class C (50 PHR)	29	57	153	501	2125

Table 5. Summary of compressive modulus properties after 3% KCl soaking and 200°F drying

To further investigate the stimuli-responsive property of fly ash filler, the latter was incorporated into a water swellable compound, which provided for significantly more interaction of filler with water to occur. Sample F2 comprises unmodified commercial water swellable NBR formulation F1 mixed with 15 wt.% of fly ash Cass C. Sample F2 was processed into a 0.04" slab, then a rectangular sample was cut from the slab and swollen in 3% KCl solution until 100% volume increase. The F2 sample was then dried overnight in the oven, which lead to de-swelling of the sample due to solvent evaporation. Mechanical properties of the F2 sample were assessed with DMA prior to the swelling test, in a swollen state, and after drying - the data summary is shown on Figure 3.

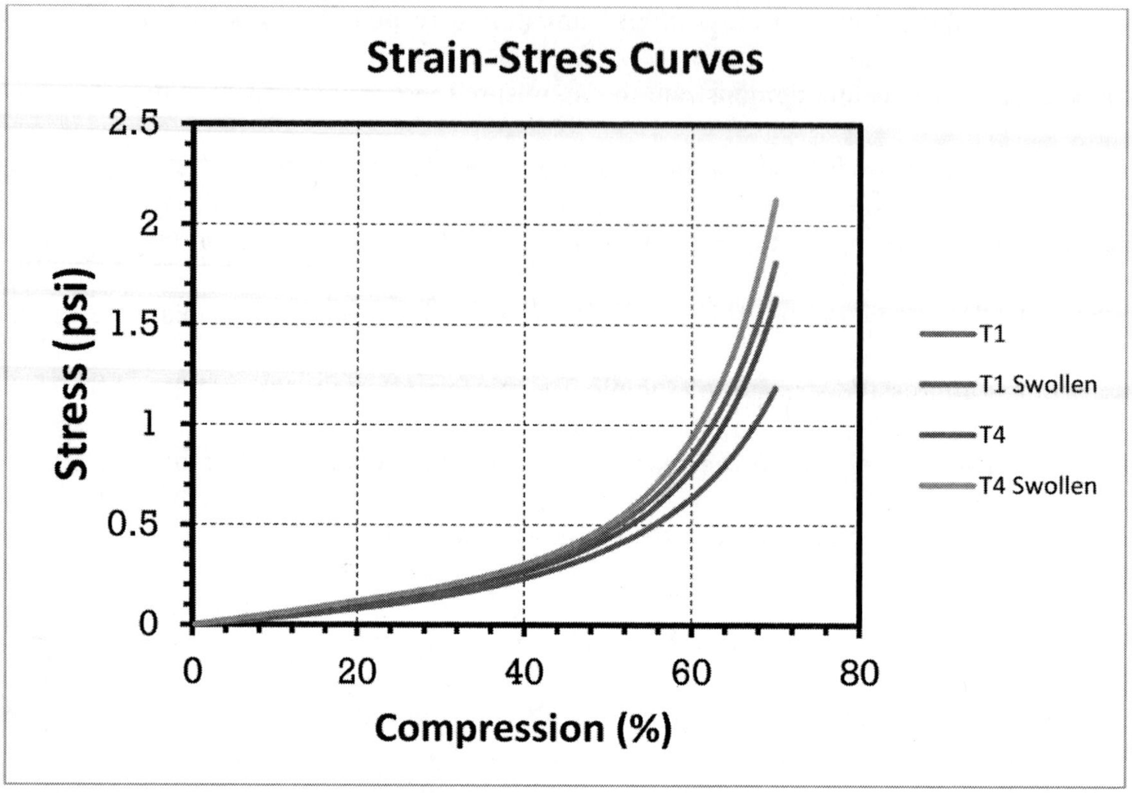

Figure 2. Stress-strain curves

The storage modulus of sample F2 after swelling and subsequent drying was about an order of magnitude higher than that of original sample F2 measured prior to swelling experiment. The storage modulus of the dried F2 sample was also less sensitive to the increasing temperature, contrary to the weakening original sample F2. The increase in the storage modulus of swollen F2 with temperature is simply due to the drying of the sample. The obtained result demonstrated that fly ash can essentially act as a stimuli-responsive material.

The significant improvement in the modulus is believed to be due to the neutralization of the swellable component of the commercial F1, and potential hydrolysis with subsequent neutralization of NBR elastomer. This is confirmed by the shift in T_g to a lower temperature range after drying as illustrated in Figure 3.

Carboxylic groups were detected in the original sample F2 prior to swelling, hence, they cannot be attributed solely to hydrolyzed HNBR, but rather to the super-absorbent polymer present in water swellable rubber compound (see Figure 4). FTIR confirms neutralization of carboxylic groups because a significant increase of the peak near ~1540 cm^{-1} and a lowering of peak intensity at ~1720 cm^{-1} have been observed. This neutralization is believed to promote the formation of ionic clusters in the compound with fly ash considered to be the source of cations. The neutralization of carboxylic groups could as well be attributed to K^+ cations present in the KCl brine solution, however, in that case, the respective peak in FTIR spectra would appear at higher wave numbers, approaching ~1600 cm^{-1}.

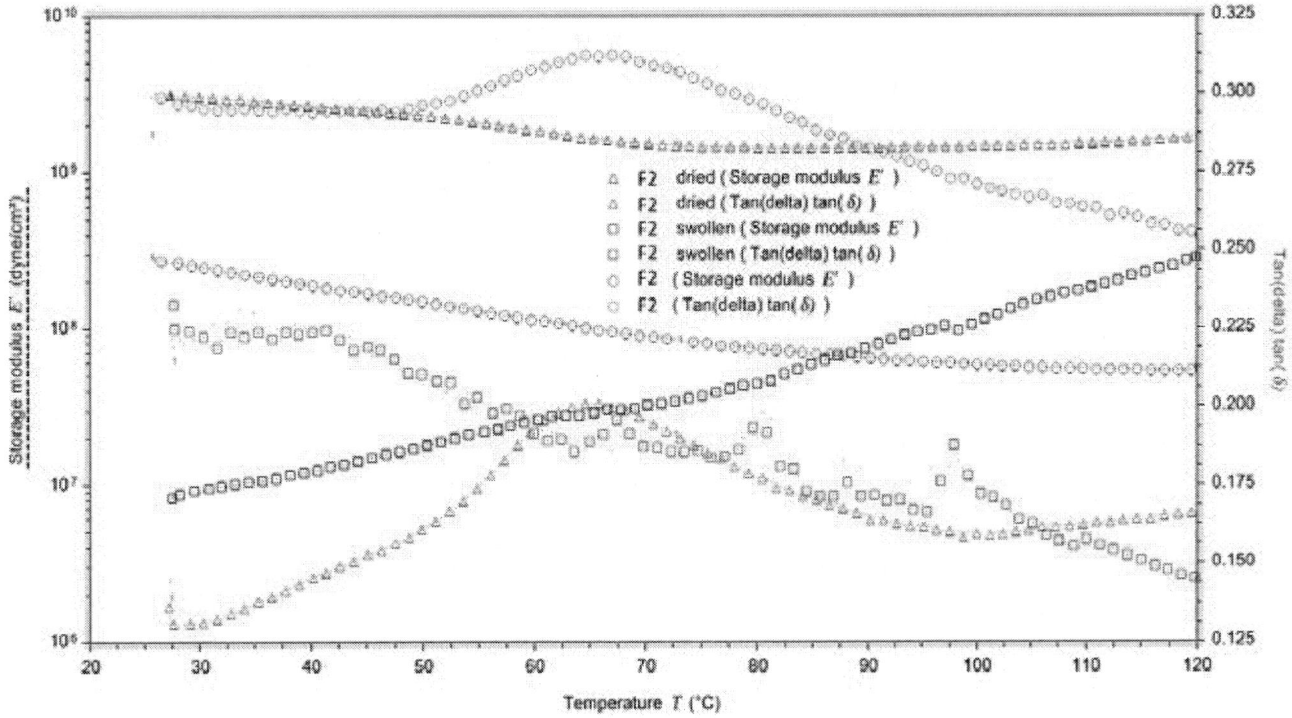

Figure 3. DMA data on F2 sample

It is important to mention that sample F2 exhibited a certain degree of brittleness after drying, which is undesired in rubbers, but this can be managed by incorporation of common ionomer plasticizers.

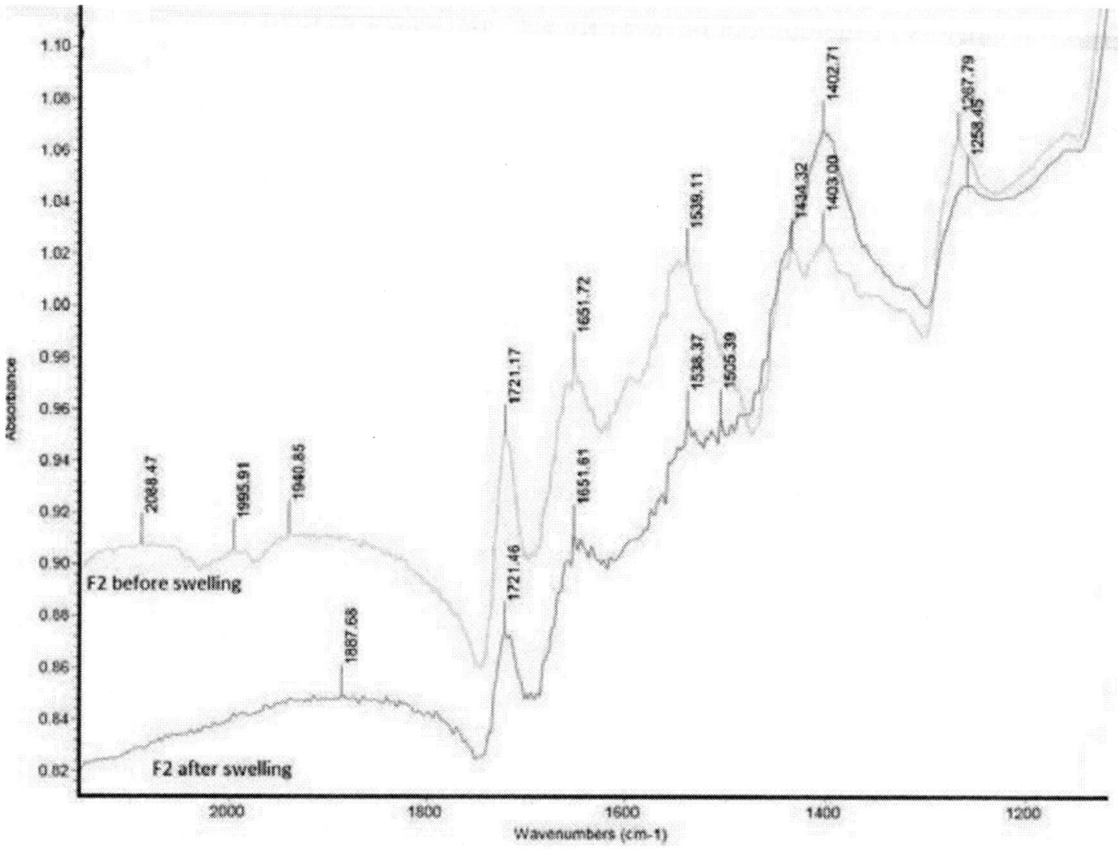

Figure 4. FTIR spectra of F2 sample taken before and after swelling

Conclusions

Fly ash demonstrated a great potential for consideration as cost effective, low reinforcing filler for downhole oil and gas applications. Although the reinforcing properties of fly ash are inferior to carbon black, incorporation of fly ash can improve the mechanical properties of HNBR elastomer compared to the unfilled raw crosslinked elastomer at both room temperature and at 325°F.

Furthermore, composites of HNBR with fly ash Class C demonstrated a stimuli-responsive effect by significantly increasing their modulus upon swelling in water brine. This effect was even more pronounced in composites of fly ash with swellable water compounds, which provided an easy access of filler to water upon swelling. A ten-fold increase in modulus was achieved in such fly ash reinforced sample, after swelling in 3% KCl water brine and drying in the oven overnight. The FTIR confirmed neutralization of carboxylic groups present in the investigated system, which contributed to overall improvement of mechanical properties of the compound upon swelling.

444093 New Development on Hoist Scheduling with the Relaxation of Cyclic Precondition

Abstract:

Multi-Stage Material Handling (MSMH) processes are designed to manufacture materials requiring a series of treatments and in industry, they are broadly used due to their capability to handle large work orders and compatibility with materials with different recipes. A typical example is electroplating process. A hoist handles the transportation of each job among units (stages) according to its processing recipe. The hoist movement plan is predetermined through optimization and this plan is repeated to perform continuity. Obviously, the ultimate goal is to maximize the productivity within each movement plan, which is equivalent to minimize the cycle time to finish one work order. The optimization process is generally called Cyclic Hoist Scheduling (CHS), since every movement plan is a cycle due to its repeatability. This precondition in term of hoist operation and unit usage is general for most cases. However, it could block the really optimal results.

In this study, this precondition has been relaxed for multi-capacity units, thus the unit usage can be more flexible. Thus, the real optimality can be found through CHS. To deal with the cycle connection issue from the relaxation of the precondition, a new module Cycle Connection Identification (CCI) is proposed. Obtained hoist schedule will be checked. Once a schedule with cycle connection problems is identified, a new cycle will be scheduled forcing the smooth connection with the first cycle. The objective of scheduling a new cycle is to match the beginning of the first cycle and the ending of the second cycle. If the two-cycle group is still with cycle connection issue, more cycles will be scheduled till the cycle group is free from cycle connection problems. Each cycle is called a sub-cycle and the constructed cycle group without cycle connection problems is called a full cycle. Note that in each sub-cycle, the work order and all transportation operations are completed.

In this paper, hoist scheduling models with/without such a precondition have been developed and proposed as mixed integer linear programming. The performance of the two models is tested through case studies and the difference is compared and demonstrated in detail.

444310 Formal Modeling of Accidents and Automatic Generation of Tutoring Scenarios for Intelligent Safety Training of Process Operators

Abstract:

With the emphasis of accident-free operations and prevention of catastrophic events, safety-training demands for the chemical plants have been on the increase globally. However, the research results on modeling of accidents and involving human operators and the advanced theory and technology for (safety) operator training have been shared limitedly among domain experts and commercial developers. Even with the recent trend of simultaneously delivering the operator training system with the construction of a new process plant, in commercial operator training systems (OTS), safety training used to implement quite limited and obvious scenarios (and the burden was almost up to the trainer), partially due to the difficulty in modeling propagations of probable accidents. By the regulatory requirements or proactive investment by companies, effective scenarios for training of safety and abnormal situation management are necessary to be improved in the aspects of the size of volume as well as the training depth. Thus, considering the emergence of enhanced virtual reality technologies and inexpensive gadgets like HMD, supporting the immersive training environment, the future efficiency of the training heavily relies on the quantity and quality of safety scenarios, which would be essential in guiding the entire training process.

In this research we first overview a government-sponsored research project developing an enhanced safety training system for chemical process operators, using haptic devices in addition to enhanced audio and visual equipment. The initial version of the system runs with a fixed set of training scenarios developed for predetermined accident scenarios. However, with a limited set of training scenarios, the trainees simply memorize and rote learn the steps, required to be mastered to proactively solve the problems with an advent of possible accidents. In addition to that, without formal modeling, the system cannot give timely feedbacks and the performance of the trainee cannot be measured in real-time. Then to enhance the contents and usefulness of the training, the knowledge of process operations is collected and systematically stored as a form of knowledge base, in addition to the accident scenarios obtained by executing Fault Tree Analysis and HAZOP study.

By using a formal modeling language like UML and Petri nets, the accident scenarios and operators' problem-solving tasks are modeled hierarchically and classified into distinguishable subtasks. Then a new training scenario is generated on-line through automatic synthesis based on planning using the knowledge base. Safety training scenarios are generated intelligently for the tutoring of a single operator's job function as well as effective collaboration among a group of operators for real-scale malfunction diagnosis and intervention at the initial propagation stage of an accident. As case study, safety trainings for city gas governing stations are executed based on systematic scenario generation and analyzed for improved effectiveness and field applicability of the system. Reduction of time and cost in the training scenario production and real-time customized training based on new operator's level of knowledge and skill are also discussed in the framework of intelligent tutoring systems.

Experimental Study on Najas Minor of Producing Bio-Oil

Abstract: Due to the diminishing of petroleum reserves and the environmental degradation caused by consumption of fossil energy , many attentions have been paid to the application of biomass energy which is acceptable as an abundant renewable energy resource. Najas minor grows in dense clusters with highly branched stems of which a small fragment could propagate into a thrive one.In this study, we researched the properties of Najas minor to make biofuels .Firstly,the najas minor after heavy metal ions degradation were fed into pyrolyser to prepare biofuels. Significant differences of the product was observed compared with only Najas minor pyrolysed.

KEY WORDS: Najas Minor; catalytic pyrolysis; bio-oil

1.Introduction

Energy demands and environmental problems caused by fossil energy have been a matter of growing concern.Bio-oil has been considered to be a promising way for producing renewable bio-fuels or chemicals[1-3].In recent years,many countries pay attention to the importance of biomass that is an abundant renewable energy resource.Biomass can be defined as any organic material mainly consisting of carbon, hydrogen,oxygen, nitrogen, and some other components in small proportions[4].Although the biomass has been studied for nearly a century,most studies mainly concentrated in Lemna minor, Lemna gibba, Spirodela polyrhiza and Landoltia punctata.We still know little about the biological characteristics and potential application value of other varieties of biomass[5].Najas minor,a submersed herb,grows in dense clusters with highly branched stems of which a small fragment could propagate into a thrive one.Najas Minor is native to Europe, Asia and North Africa from the Netherlands to Morocco east to Japan and the Philippines,including China,Siberia, Central Asia, Iran, Turkey, Ukraine, Germany, France Italy and a host of other Countries.

Catalysts used for catalytic cracking are basically divided into the following categories: molecular sieve, alkali metal salt, metal oxide, etc.Different catalysts have different catalytic effects.The use of metal nanoparticles has shown to be very important in recent industrial applications[6].The application of precious metal nanometer materials in catalysis,which has caused the extensive concern of researchers in recent 30 years,has become a hot research topic in the field of catalysis.Gold nanoparticles have received a great deal of attentions due to their special optical and electrical performances, catalytic properties and remarkable quantum size effect, surface effect etc[7]. The synthesis of metal nanoparticles can be classified broadly into wet and dry methods[8]. The cost of production is relatively high both materially and environmentally. There is an unequivocal need to develop more cost-effective and environmental benign methods to alternate these existing methods[9].Researchers in this field turn to use organisms to produce inorganic nanostructures extracellularly[10].

Although there has been a large volume of work on the synthesis of gold nanoparticles[11-13], and obtainning bi-oil by abundant biomass, there are relatively fewer attempts to combine these two methods.In this work, gold nanoparticles were obtained by aqueous chloroauric acid solution with the of Najas Minor at room temperature. The direct pyrolysis and catalytic pyrolysis experiments of Najas Minor were carried out to observe the effects of catalysts on pyrolysis process of Najas Minor.

2. Materials and Methods

2.1. Experimental samples

The Najas Minor collected from Tianjin city and the solutions of Au(III) were used in this study.The stock solutions of Au(III) were prepared by dissolving analytical grade chloroauric acid

($AuCl_3 \cdot HCl \cdot 3H_2O$, Kewei Reagent, China) in deionized water.

2.2.Experimental apparatus and procedure

Pyrolyzer with GCMS,shown in Fig. 1,is mainly used for analyzing the composition of polymer.Pyrolyzer with GCMS includes three parts :quartz pyrolysis tube,pyrolyzer and GCMS.

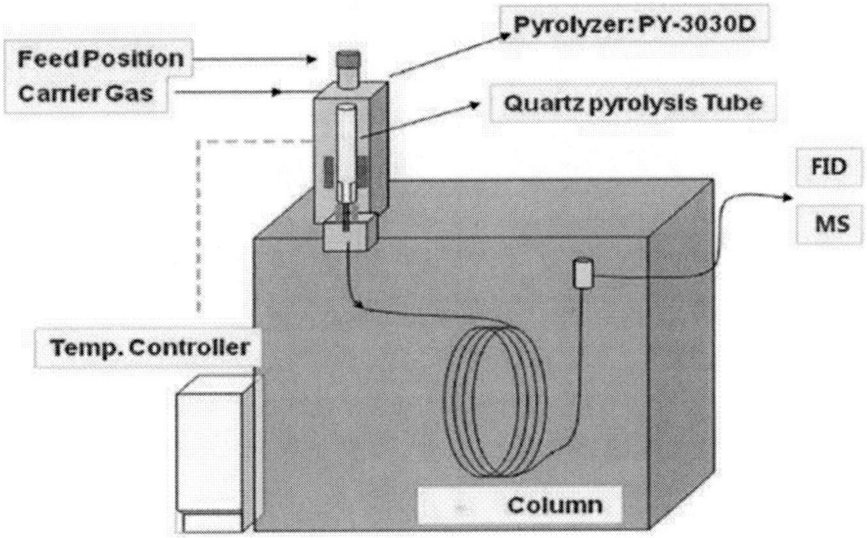

Fig. 1 Diagram of Pyrolyzer with GCMS

The quartz pyrolysis tube (shown in Fig.2) length is 12cm.The sample end diameter is 5 mm,and the pyrolysis products end diameter is 3mm.This special structure is beneficial to the fixation of the reactant in the tube.The top and bottom of the reactants are both supported by quartz wool to prevent reactants blowed by the carrier gas ,or fine particles into the chromatographic column.All pyrolysis experiments were carried out on the EGA/PY-3030D pyrolyzer(Frontier Lab,Japan) .The samples were filled in the special quartz pyrolysis tube to reduce the amount of valuable catalyst and reduce the cost of research work.

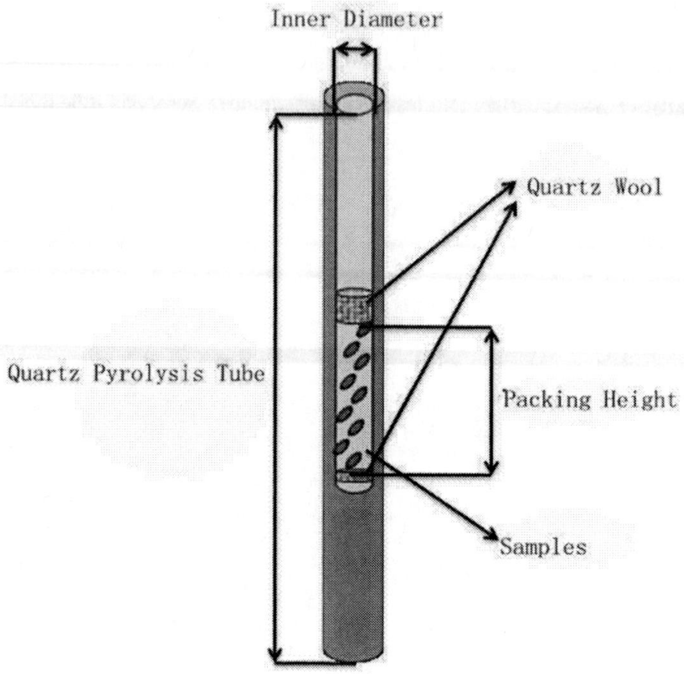

Fig. 2 Diagram of Quartz Pyrolysis Tube

Approximately 0.5g Najas Minors were used in each experiment.The samples were impregnated in 50ml deionized water and 50ml solution of chloroauric acid(0.0793mmol/L) for a night,followed by drying at 80℃ until the weight didn't change.Each experiment used 2.5mg Najas Minors as reactants.Each sample was heated, under the control of heating controller, from ambient temperature to 600 ℃ repeatly and kept at 600 ℃ for 2 h.

3. Results and discussion

3.1.MS Analysis of the Products during the Pyrolysis

Under the same pyrolysis conditions, bio-oil obtained by direct pyrolysis and catalytic pyrolysis were analyzed by GCMS,and the main chemical compounds were shown in Table1 and Table2.

Table 1 Indentification and quantification of main chemical compounds by direct pyrolysis

NO.	peak area%	Formula	Compound
1	12.67	$C_3H_6O_2$	2-Propanone, 1-hydroxy-
2	3.67	C_7H_8	Toluene
3	3.28	C_7H_8O	p-Cresol
4	3.81	$C_5H_{10}O$	Pentanal
5	3.53	C_8H_8O	Benzofuran, 2,3-dihydro-
6	2.17	C_8H_8N	Indole
7	5.93	$C_{20}H_{40}O$	3,7,11,15-Tetramethyl-2-hexadecen-1-ol
8	2.52	$C_{22}H_{42}O_2$	Phytol, acetate
9	5.25	$C_{16}H_{32}O_2$	n-Hexadecanoic acid
10	8.99	$C_{18}H_{30}O_2$	9,12,15-Octadecatrienoic acid, (Z,Z,Z)-

Table 2Indentification and quantification of main chemical compounds by catalytic pyrolysis

NO.	peak area%	Formula	Compound
1	6.01	C_7H_8O	p-Cresol
2	1.75	$C_7H_7NO_2$	3-Hydroxypyridine monoacetate
3	4.58	C_8H_7N	Indole
4	3.05	$C_{10}H9N$	1-Phenyl-1-cyclopropanecarbonitrile
5	3.90	$C_{12}H_{22}O_{11}$	.alpha.-D-Glucopyranose, 4-O-.beta.-D-galactopyranosyl-
6	8.04	$C_{20}H_{40}0$	3,7,11,15-Tetramethyl-2-hexadecen-1-ol
7	2.82	$C_{20}H_{40}$	2-Hexadecene, 3,7,11,15-tetramethyl-, [R-[R*,R*-(E)]]-
8	3.99	$C_{15}H_{28}$	1-Pentadecyne
9	2.12	$C_{15}H_{28}$	3,7,11-Trimethyl-2,4-dodecadiene
10	8.75	$C_{16}H_{32}\ 0_2$	n-Hexadecanoic acid

3.2 discussing

According to the chemical category and common characteristics, the products were divided into hydrocarbons, alcohols, acids, ketones, nitrogen compounds and other compounds.Nitrogen compounds detected by GC/MS analysis ,mainly come from the pyrolysis of amino acids in Najas Minor.Table 3 showed the distribution of pyrolysis products.As shown in the table,the hydrocarbons ,alcohols and nitrogen compounds in biological oil of catalytic pyrolysis were significantly increased compared with direct pyrolysis.The content of hydrocarbons increased from 12.4% to 34.6%.The content of alcohols increased from 9.1% to 16.6%.The gold nanoparticles loaded in surface of Najas Minor, was conducive to the generation of hydrocarbon and alcohol in pyrolysis.By analyzing the GC/MS results, the main hydrocarbons in bio-oil by catalytic pyrolysis were long chain alkanes,and olefins.The main components in gasoline and diesel is hydrocarbons. As a result, whether the biomass can be effective as future power fuel oil, depends largely on the hydrocarbon content of bio-oil after further treatment.

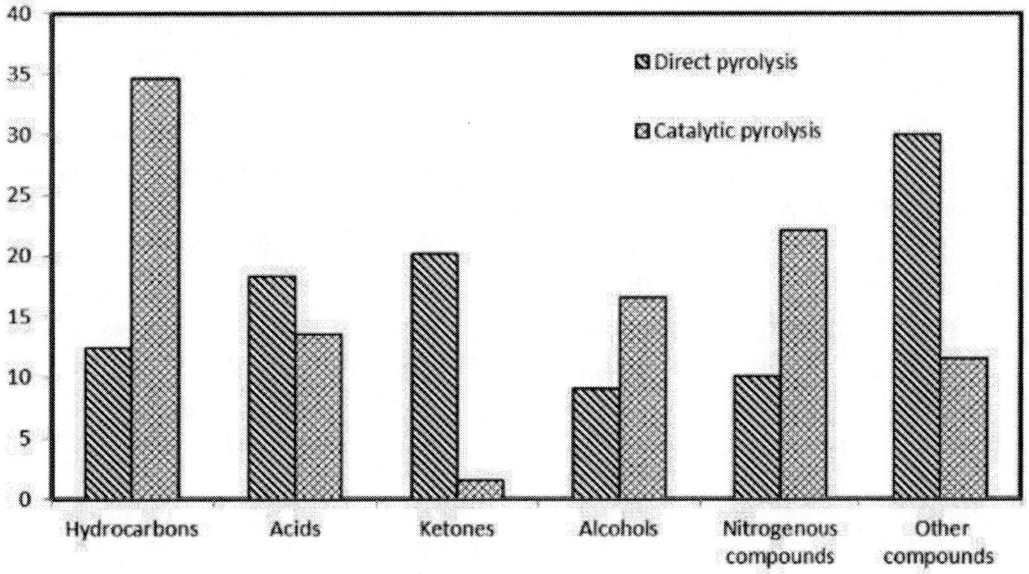

Fig.3 Distribution of pyrolysis products

4.Conclusion

Biomass was explored as a potential biomass energy resource. The experimental study on the direct and catalytic pyrolysis of Najas Minor had been presented, using Pyrolyzer with GCMS,to observe the catalytic effects of gold nanoparticles.In this study,remarkable catalytic effects for pyrolysis were observed.The pyrolysis of Najas Minor with gold nanoparticles,provided more hydrocarbons, alcohols and nitrogen compounds.

参考文献：

[1]Gong F, Yang Z, Hong C, et al. Selective conversion of bio-oil to light olefins: Controlling catalytic cracking for maximum olefins[J]. Bioresource Technology, 2011, 102(19):9247-54.

[2]Zhang Z, Bi P, Jiang P, et al. Production of gasoline fraction from bio-oil under atmospheric conditions by an integrated catalytic transformation process[J]. Energy, 2015, 90:1922-1930.

[3]Hui S, Yuqi D, Jinzhao D, et al. Transesterification of sunflower oil to biodiesel on ZrO2 supported La2O3 catalyst.[J]. Bioresource Technology, 2010, 101(3):953–958.

[4]Yaman S. Pyrolysis of biomass to produce fuels and chemical feedstocks[J]. Cheminform, 2004, 35(31):651-671.

[5]Hejný S. Ellias Landolt The family of Lemnaceae—A monographic study—Vol.1 biosystematic investigations in the family of Duckweeds (Lemnaceae) (vol. 2)[J]. Folia Geobotanica Et Phytotaxonomica, 1993, 28(1):50-50.

[6]Gardea-Torresdey J L, Tiemann K J, Gamez G, et al. Gold Nanoparticles Obtained by Bio-precipitation from Gold(III) Solutions[J]. Journal of Nanoparticle Research, 1999, 1(3):397-404.

[7]Marie-Christine D, Didier A. Gold nanoparticles: assembly, supramolecular chemistry, quantum-size-related properties, and applications toward biology, catalysis, and nanotechnology.[J]. Chemical Reviews, 2004, 35(16):293-346.

[8]Liu B ,, Xie J ,, Lee J Y, et al. Optimization of high-yield biological synthesis of single-crystalline gold nanoplates.[J]. Journal of Physical Chemistry B, 2005, 109(45):15256-63.

[9]Raveendran P, Fu J, Wallen S L. Completely "Green" Synthesis and Stabilization of Metal Nanoparticles[J]. Journal of the American Chemical Society Jacs, 2003, 125(46):págs. 13940-13941.

[10]Absar Ahmad , Satyajyoti Senapati , MIK, et al. Extracellular Biosynthesis of Monodisperse Gold Nanoparticles by a Novel Extremophilic Actinomycete, Thermomonospora sp.[J]. Langmuir, 2003, 19(8):3550-3553.

[11]Selvakannan P, Mandal S, Phadtare S, et al. Water-dispersible tryptophan-protected gold nanoparticles prepared by the spontaneous reduction of aqueous chloroaurate ions by the amino acid[J]. Journal of Colloid & Interface Science, 2004, 269(1):97-102.

[12]Gittins D I, Caruso F. Biological and Physical Applications of Water-Based Metal Nanoparticles Synthesised in Organic Solution[J]. Chemphyschem, 2002, 3(1):110－113.

[13]And A G, Murphy C J. Seed-Mediated Synthesis of Gold Nanorods: Role of the Size and Nature of the Seed[J]. Chemistry of Materials, 2004, 16(19):3633-3640.

COMPARATIVE ANALYSIS OF MOLECULAR STRUCTURE IDENTIFIABILTY BASED ON SIGNATURES AND DESCRIPTORS

Zelimir F. Kurtanjek
University of Zagreb, Croatia

Abstract

This work is focused on molecule computer representation aimed for molecule predictive model (QSAR) and inverse modelling. Two different approaches are compared, i.e. the molecule structure coding based on graph theory (Faulon et al.[1] extended valence) and the chemical based descriptors[2]. Applied are 1-3rd level molecule graph fragments and 630 2D molecular descriptors. The molecule test sets are alkanes, alkenes, acetones, aromatics, organic acids and halogenated hydrocarbons in the range of C_1-C_{12}, selection of biomolecules (including aminoacids) and a set of binary ionic liquids (cations: imidazole, pyridinium, quinolinium, ammonium, phosphonium). The comparative comparison of molecule description based prediction potential are statistically evaluated by octanol-water (PWA) partition coefficients (for hydrocarbons), viscosity of ionic liquids and biological impact on acetycholinesterase inhibition. The evaluation is based on two distinct levels. On the first level the comparative analysis based on internal and external correlation between "scores" in multidimensional spaces of the separate first, second and third order valance spaces, and the molecular signature space. Firstly, the potential of information content applicable for inverse modeling is tested on ability for molecule classification by use of linear discrimination (LDA) based on the singular value decomposition (SVD) of corresponding molecular fragments and descriptors. For chemical-biological property inverse modelling proposed is application of constrained genetic algorithm (GA). Tested are linear structure multivariate (chemometric) models such as principal component regression and partial least squares and nonlinear decision trees (random forest). Applied are singular value decompositions for reduction of the molecule signature and the descriptor spaces up to the significance level of p=0.05. The selected level of significance is presumed to be in coherence with experimental data of molecule physical and chemical properties. The main focus is on extraction of the main molecular fragments (partial graphs extracted from molecular signatures) and specific molecule descriptors. Importance of application of open source software modules provided by Bioclipse[3] and R software packeges[4] for computational-statistical numerical algorithms is emphasized.

Key words: Molecule signatures, descriptors, inverse modelling, decision tree

Introduction

Advancement of high-throughput experiments have lead to „big data" databases applied in pharmaceuticals, medicine, new material science, catalysis, and new industrial processes. The role of chemoinformatics as a science for computer encoding of molecule properties and algorithms for data mining and inference are the essential steps in the development. Computer encoding of molecular properties has been a part of chemical engineering from the beginning of process design. It is mostly based on molecule group contributions, but has been extended by application of molecular descriptors (1-2-3 D), molecule signatures, and application of molecular quantum mechanics. Modeling is faced with the problem of „high dimension curse". For example, about 5 000 various molecular descriptors can be determined for each molecule, although most of them are highly correlated. It results in an ill-conditioned problem of modelling with large uncertainties in parameter and poor model prediction accuracy outside modelling training sets. Great improvement has been achieved by application by use of model ensembles, like random forest of many individual decision trees. These models are highly flexible, can be trained on discontinuous and nonlinear functional relations, can encompass various data formats like numerical and categorical variables, and due to large number of randomly selected individual predictors in principle do not have tendency to overfitting. However, when the object of modelling is extraction of important molecular features, due to autocorrelation between descriptors random forest, produced lists of importance are large sets of poorly distinguishable variables. The focus of this work is to apply feature regularization for optimization and reduction of list of variable importance. Compared are two methods for structure encoding, molecular descriptors and molecular signatures of first and second order.

Modeling and Factor Regularization

In this work are selected two distinct data sets of organic molecules from open source data bases (J. Sangster, 1989) and ionic liquids ILs (The UFT/ Merck and NIST, 2015), and lists of the selected molecules are given in Table 1. As target properties for organic molecules are selected the corresponding molecular classes and partition octanol-water (POW) coefficients, and dynamic viscosity in the range of 30-80 ^{0}C of ILs is selected The molecules are coded into SMILES and used as input lists into software for evaluation of molecular signatures. Most popular open source software tools are DRAGON (R. Todeschini, et al., 2009) and PADEL (Chun Wei Yap, 2010). Here are used 1-2 D descriptors generated by PADEL. For the same list of molecules are determined first and second order molecular signatures or extended valances (J.P. Faulon, 2003). The signatures are two dimensional graph based structural molecular fragments. The vertices are atoms and the edges are corresponding bonds (lengths or heights), and are here only used of the first and second level. A molecule is characterized by a complete set of signatures as input data for modelling. Since the models are based on large number of variables (molecular features) possibly lager than number of samples (molecules), usually denoted as p>>n models. They are intrinsically „ill conditioned" leading to good predictions on train data but very poor predictions on test data. Also, produced list of variable importance are very dense "clusters are not distinguished" with large uncertainties (broad confidence intervals). This makes the problem of inverse modelling aimed to predict a set of candidate molecules for given set of target properties, such as in a case of high dimension combinatorial task for selection of ILs cations and anions, very unreliable. In order to reduce model dimensionality and improve robustness and parsimonious of important molecular features here is applied modelling with feature regularization. The model is based on ensemble of random trees (CART models, L. Breiman et al. 1998) with included feature regularization.

Data sets of organic molecules		Range of number of atoms			
Molecular classes	C	H	O	Cl	F – I - Br
Alkanes	5-12	12-30			
Alkenes	4-9	6-18			
Acetones	6-24	6-26			
Halogenated hydrocarbons	3-10	6-22	1		
Organic acids	1-16	2-32	2		
Esters	4-11	6-22	2		
Aromatics	1-12	1-21		3	2 – 2 - 1
Data sets of ionic liquids (ILs)					
Cations		**Anions**			
1,3-Dimethylimidazolium (+)		PF_6 (-) Cl (-) CH_3COO (-)			
1-Methylpyridinium (+)		BF_4 (-) $MeSO_4$ (-) $EtSO_4$ (-)			
1,1-Dimethylpyrrolidinium (+)		Tf_2N (-) CF_3SO_3 (-)			

Table 1. The molecular groups of the organic and ionic liquid molecules selected for random forest models and extraction of the most important molecular features. The properties for the organics are obtained from J. Sangster (1989) and for the ionic liquids from Merck and NIST data bases (2015). Molecular descriptors and signatures of ILs ions are recorded separately and adjoined as a single record.

The modelling is formulated as an optimization (functional and multivariable minimization) problem with two parts of the objective function (T. Hastie et al., 2009).

$$\min F_{object.}(X,T,\omega) = SSE(y,\hat{y}) + \phi(T,\omega) \tag{1}$$

The first part is the sum of square errors (SSE) based on difference between observed y and prediction $\hat{y}$ based on input data by an ensemble T of interrelated CART models (L. Breiman et al. 1998), The second part is a selected metrics ϕ for the model complexity depending the sum of leaf probabilities for population of individual CART models and their total number N_T .

$$\phi(T,\omega) = \gamma \times N_T + \frac{1}{2} \times \lambda \times \sum_{i}^{N_T} \omega_i^2 \tag{2}$$

The ensemble output prediction is a sum of predictions of individual T_i tree CART models:

$$\hat{y} = \sum_{i}^{N_T} T_i(X) \tag{3}$$

A very efficient open source code Extreme Gradient Boosting XGB (T. Chen, 2016) is available in R language (R Developm-ent Core Team, v. 3.2.3, 2015). The optimization algorithm is consecutive, starting with first CART model which is appended with a next tree boosted by errors in predictions of previous tree. At each step optimization is solved by linearization of the objective function to quadratic terms. Number and structure of trees as well as the parameters for regularization need to be found by trial and error and minimal number of trees by folding and cross validation error on the training set. The critical step is validation on untrained set to have acceptable model error and avoid model overfitting with minimal model complexity.

Results and Discussion

Tested are models for prediction of the seven molecular classes (Table 1.) based on molecular descriptors and molecular signatures (first and second level). Results for molecular descriptor feature importance are shown on Fig. 1. The effect of regularization is very pronounced. Out of hundreds of descriptors for the molecular classification needed are only few, the most important being BCUTw.1h. On contrary, when an ensemble of independent CART trees is applied, i.e. random forest model, level of accuracy of prediction is the same, but the list of variable importance is densely populated with numerous descriptors.

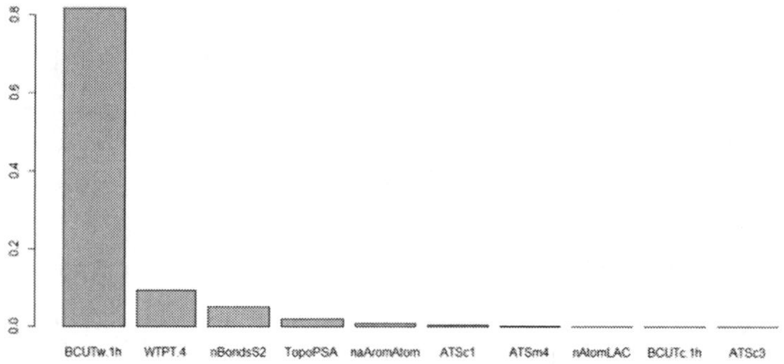

Fig 1. Variable importance molecular descriptor factors for the inverse modeling of the selected molecular classes. The first five descriptors are: BCUTw.1h (based on a weighted version of the Burden matrix which takes into account both the connectivity as well as atomic properties of a molecule), WTPT.4 (Randić's weighted path), nBondS2 (bond count), and naAromAtom (number of aromatic atoms).

The same classification problem was tested by the first and second order molecular signatures. The important signatures are depicted on Fig. 2-3. Accuracy of prediction on trained sets of molecules compared to descriptors is about the same, though the importance factors are not as distinctive as in case of the descriptors.

significantly better when molecular descriptors are applied. The

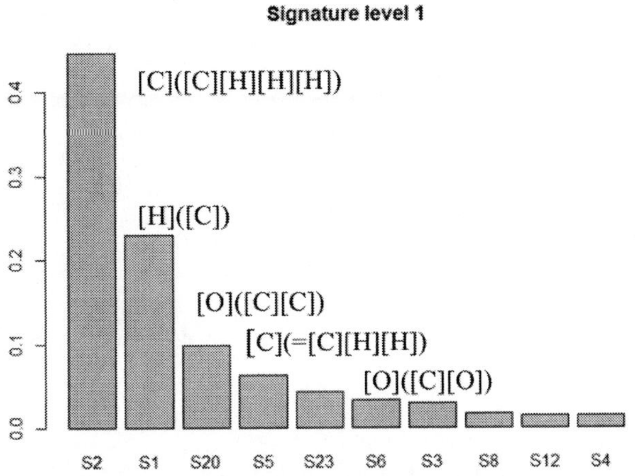

Fig 2. Relative importance of the individual first level signatures for inverse prediction of the selected molecular classes. The first five most important signatures are coded according to SMILES (Simplified Molecular Input Line Entry System).

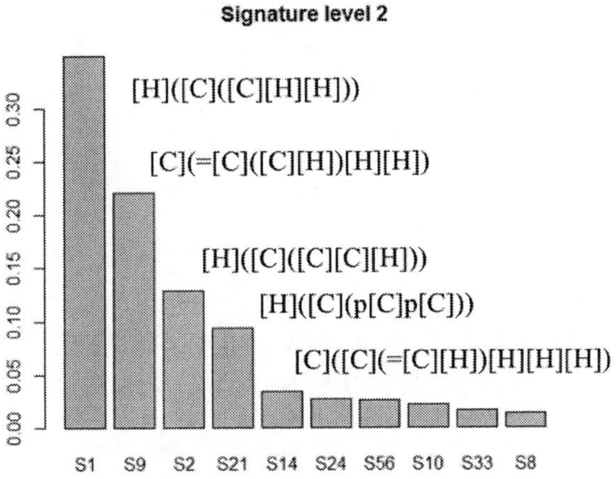

Fig 3. Relative importance of the individual second level signatures for inverse prediction for the selected molecular classes. The first five most important signatures are coded according to SMILES (Simplified Molecular Input Line Entry System).

However, model accuracy on untrained set of molecules, depicted in Fig. 4, is significantly lower for the model with molecular descriptors compared to signatures. The average error with descriptors is about 5% while for signatures about 11%. The "box and whiskers" plots are obtained by error evaluation of ten randomly generated untrained sets.

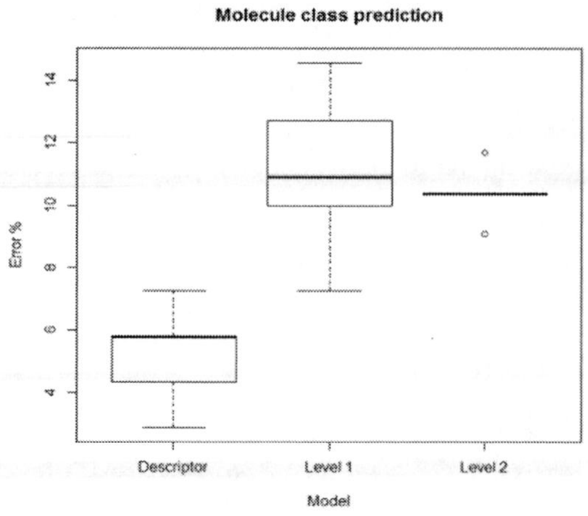

Fig 4. "Box and whiskers" plot for the model validation (relative errors) for inverse prediction of the molecular classes based on the set of untrained molecular descriptors and the first and second level of molecular signatures.

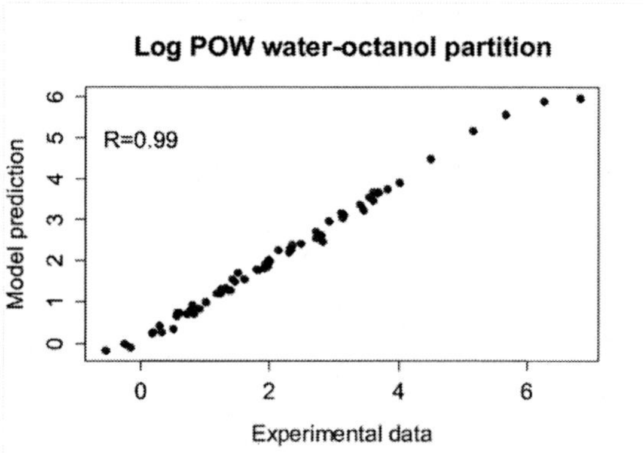

Fig 5. Predictions of the partition octanol-water coefficients POW, for the specified molecular classes of organic molecules, are based on the random forest model with the molecular descriptors.

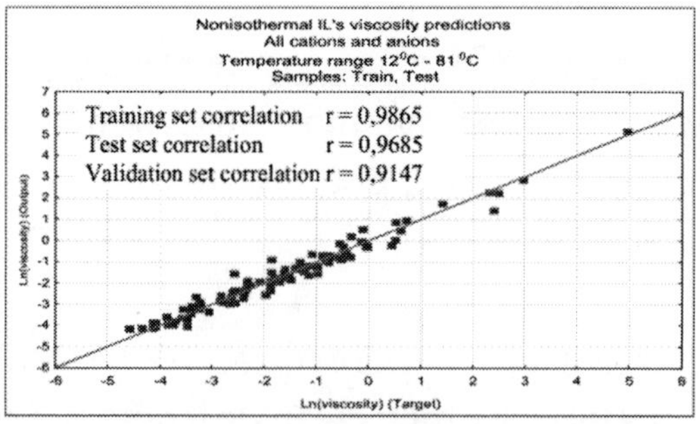

Fig 6. Predictions of dynamic viscosities for the specified groups of ionic liquids (ILs) based on the random forest model with the molecular descriptors. The viscosities cover the temperature range from 12 – 81 ^{0}C and five orders of magnitude.

In Fig. 4 can be observed that dispersion of errors (quantiles) for the second level molecular signatures is negligible, though two outliers are present.

For prediction of molecular properties applied are random forest models based on molecular descriptors. For all of the organic molecules, included in the seven classes, evaluated are octanol-water partition coefficients POW, shown in Fig. 5. The range of POW values is very high, on log scale fro 0 to 6, but the obtained accuracy of the model predictions is high (Pearson correlation R=0.99). The same approach is applied for prediction of ionic liquid viscosity data in temperature range from 12 to 81 ^{0}C. Since temperature effects on viscosity are intrinsically nonlinear, they represent a modelling challenge. However, the decision trees are very efficient for accounting nonlinearities, though nonlinearity is reflected in non-orthogonal decision planes, rendering trees with significantly increased number of branches. For the range of 6 order of magnitude the accuracy of prediction on trained set is R = 0.98 and on untrained set R = 0.91.

Conclusions

Compared are ensemble decision trees models for molecular feature class and property prediction based on extensive list of molecular descriptors and first and second level of extended valances or signatures.

The aim was to test effects of regularization on ensemble of CART decision tree models. Proven is effectiveness of the extreme gradient boosting algorithm in cases with extensive list of (molecular) features.

Obtained is increased accuracy on untrained sets of data, and more importantly, obtained are parsimonious lists for ranking of variable importance lists. This enables further significant reduction in development of robust and simple decision tree models.

The reduction in number of molecular features applied for decision tree models may greatly simplify the challenging problem of inverse modeling based on molecular descriptors and/or signatures.

References

1. J. Sangster, J. Phys. Chem. Ref. Data, **18**(3)(1989) 1111-1227
2. The UFT/ Merck Ionic Liquids Biological Effects Database, www.il-eco.uft.uni-bremen.de
3. NIST, Ionic Liquids Database Standard Reference Database #147, <ilthermo.bouder.-nist.gov /ILThermo/mainmenu.uix>
4. R. Todeschini and V.Consonni: "Molecular Descriptors for Chemoinformatics", (2 volumes), WILEY-VCH,Weinheim (Germany) 2009.
5. Chun Wei Yap, PaDEL-Descriptor: An Open Source Software to Calculate Molecular Descriptors and Fingerprints, J. Comp. Chem., (2010):**32**: 1466-1474.

6. J.-L. Faulon, D.P. Visco Jr., R.S. Pophale, The signature molecular descriptor. 1. Extended valence sequences vs. topological indices in QSAR and QSPR studies, J. Chem. Inf. Comput. Sci. **43** (2003) 707–720.
7. J.-L. Faulon, C.J. Churchwell, J.D.P. Visco, The signature molecular descriptor. 2. Enumerating molecules from their extended valence sequence, J. Chem. Inf. Comput. Sci. **43** (2003) 721–734.
8. L. Breiman, J. H. Friedman, R. A. Olshen, C.J. Stone, „Classification and Regression Trees", CRC, Boca Raton, 1998.
9. T. Hastie, R. Tibshirani, J. Friedman, The Elements of Statistical Learning, Data Mining, Inference and Prediction, Springer Verlag, Second Edition, New York, 2009.
10. T. Chen, T. He and M. Benesty (2016). xgboost: Extreme Gradient Boosting. R package version 0.4-3. https://CRAN.R-project.org/package=xgboost
11. R Development Core Team, , R: A language and environment for statistical computing. R, Vienna, Austria,www.R-project.org 2015.

444478 Biogas Reforming with Addition of Carbon Dioxide to Syngas over Ni-Based Bimodal Pore Catalyst

Abstract:

Methane dry reforming of (MDR) has been widely studied due to its important applications in producing of syngas ($CO + H_2$), removal of two greenhouse gases, and upgrading of biogas (mainly composed of CH_2 and CO_2) into value added chemicals. Compared to the noble metal catalysts, nickel-based catalysts are suitable for industrial scaled-up MDR due to their high activity and low cost. However, nickel-based catalysts have been encountering a significant challenge that is the active metal sintering and coke deposition, leading to the catalyst deactivation and then the poor stability. Nevertheless, it has been reported that the deactivation of nickel catalysts can be suppressed by adding promoters like strong Lewis bases with the enhancement of chemisorb CO_2, such as MgO and CaO, and like lanthanide elements with the capacity of oxygen storage and release, such as CeO_2 and La_2O_3. Al_2O_3 is a commonly applied catalyst carrier own to its good pore size dispersion, high specific surface area and high mechanic strength.

Herein, we report that the bimodal NiCeMgAl catalysts with different NiO-loading were synthesized by the refluxed co-precipitation method and were evaluated for MDR performance by using real biogas from anaerobic digestion of food wastes. The sample containing 15wt% NiO-loading ($Ni_{15}CeMgAl$) was found to be active enough at 750 °C with a high CH_4 conversion of around 90%. $Ni_{15}CeMgAl$ catalysts kept their bimodal porosity after reduction and MDR process. The evolution of the $Ni_{15}CeMgAl$ catalyst before and after MDR process was investigated by SEM, TEM, XRD, and TGA techniques. The correlation between the structure evolution and catalytic performance change was also discussed.

444994 Multiscale Simulation of Proteins Motion in the Microchannel

Abstract:

Protein adsorption is an important phenomenon, which plays a prominent role in chemical, industrial and medical areas. In this work, the dynamic behavior of the egg-white lysozyme has been investigated in a micro-channel with a multiscale modeling approach. The Molecular Dynamic simulation, free energy computation and Langevion Dynamics (LD) simulations have been combined to study lysozyme adsorption process on Au (111) surface. The gold surface has strong attraction to proteins. The protein-surface adsorption or desorption free energies is calculated. From the case studies of LD simulations, the surface has effects on the protein dynamic adsorption process. By the comparison of LD, MD and the experiments, the time step is concluded as an important factor to study protein adsorption behavior at the water-surface interface. Fine time step at the scale of 1 picosecond is necessary to provide enough accuracy to resolve proteins' dynamics trajectories.

444827 Syngas to Ethylene Glycol: A Technology Overview

Abstract:

SABIC is the world largest MEG producer petrochemical company and most of its MEG is being produced through the hydrolysis of EO. Ethylene oxide (EO) is a simplest form of epoxides and due to its molecular structure, acts as a versatile intermediate for a variety of chemicals such as; monoethylene glycol (MEG), di- and triethylene glycols, ethoxylates, ethanol amines, and glycol ethers. The growing demand of MEG owing to its downstream utilization in polyester industry has prompted the research of an efficient, cost effective, and environmentally benign EO production process.

The major feedstock for MEG are ethylene and syngas. The heavy dependence of MEG producers on ethylene and its scarcity in some parts of the world promoted the need of alternate feedstock for highly demanded MEG. The technological developments utilizing syngas as a feedstock will be discussed along with the commercial analysis and future growth in EO/EG business.

445029 Electrospun Nanocomposite Fibers for the Removal of Heavy Metals in Water

Abstract:

Uniform electrospun nanofibers with small diameters were successfully prepared by systematically optimizing the electrospinning parameters including polymer concentration, salt loading, applied voltage, tip-to-collector distance, and flow rate. These parameters are shown to affect the fiber morphology and the addition of iron (III) nitrate salt was found to be able to reduce the fiber size. Due to the conductive nature of iron salt in polymer solution, even a relative lower voltage of 6 kV could generate a stable fluid jet and yield the nanofibers. The thermal properties of the composite fibers were investigated by thermal gravimetric analysis (TGA) and differential scanning calorimetry (DSC). Rheological studies were also conducted. The composite nanofibers were further carbonized to carbon nanofibers. Applications of the nanofibers in environment remediation have shown promise.

445004 Grand Challenges in Future Energy and Chemicals: Climate Change and Sustainability As Drivers for New Technology and Innovation

Abstract:

Global demand for energy will double by 2050, driven by increasing population and affluence especially in emerging economies. While fossil fuels will continue to underpin the energy supply for the near future, concern over impact on climate change continues to grow, driving interest in expanded use of renewables and development of carbon capture and sequestration or re-use solutions. In addition to CO_2, water use and air emissions are critical issues to be addressed in future and current energy supply, as is energy storage for effective use of renewables. This presentation will provide an overview of what industry is doing today to address these issues, including the factors and metrics used in decision making.

444825 Significance of Energy Intensity and GHG Emission Reduction Towards Sustainability of a Petrochemical Plant

Abstract:

Sustainability is an important pillar in today's petrochemical industry for improving process efficacy by reducing specific energy consumption, greenhouse gas (GHG) emissions, and wastewater. SABIC (Saudi Basic Industries Corporation) like other giant petrochemical companies has set stringent sustainability targets and investing huge resources to transform its plants from the present position to the first quartile in each category.

SABIC is the world largest producer of ethylene oxide and ethylene glycol. For each ton of glycol produced, 0.2 to 0.5 tons of CO_2 is emitted as GHG. The significance of sustainability "in general" while the methodology adopted to reduce energy intensity, carbon footprint, and wastewater "in particular", will be discussed thoroughly in our talk.

444799 Biomass Conversion to Isoprene

Abstract:

Isoprene is an important raw material in the process industry. It is a feedstock for poly-isoprene and thermoplastics. Polyisoprene is used in a wide variety of rubber applications including medical equipment, baby bottle teats/nipples, toys, shoe soles, tires, latex®, and elastic films and threads for golf balls or textiles. It is also used in adhesives, paints and coatings. Butyl rubber, co-polymer of isoprene with isobutene, has outstanding impermeability to gases and is used, for example, in inner tubes. Styrene-isoprene-rubber is a copolymer that is used in pressure sensitive adhesives. Currently, it is co-produced in ethylene cracker (from light naphtha) or pyrolytic gasoline production processes. Both these processes are fossil based.

The authors will present a novel method to convert biomass to renewable isoprene. The method utilizes metabolically engineered E. coil and low temperature post-processing to selectively produce isoprene. The process is based upon advanced fermentation. Results for conversion of biomass derived sugars will be presented. Overall mass and energy balances and process system analysis alongwith preliminary economics will be described. Ability to produce isoprene from biomass will enable renewable rubber industry and reduce the environmental impact by reducing use of fossil feedstock.

442109 Cellulosic Ethanol Production through Chemical Detoxification

Abstract:

Cellulosic ethanol production from lignocellulosic biomass offers a sustainable solution for transition from fossil based fuels to renewable alternatives. However, a few long-standing technical challenges remain to be addressed in the development of an economically viable fermentation process from lignocellulose. Such challenges include the needs to improve yeast tolerance to toxic inhibitory compounds and to achieve high fermentation efficiency with minimum detoxification steps after a pretreatment of the biomass. We developed an *in-situ* detoxification strategy by exo-protection of an industrial dry yeast (starch-base) with polymeric materials that are compatible with the yeast. The exo-protected yeast cells displayed remarkably boosted vitality with high tolerance to toxic inhibitory compounds such as phenolic and furanic compounds, and with largely improved ethanol productivity from crude lignocellulosic hydrolysates derived from pretreated lignocellulose. The chemical exo-protection makes the industrial S. *cerevisiae* yeast directly applicable for the production of cellulosic ethanol with substantially improved productivity and yield, without of the need to use genetically modified microorganisms.

444481 A Novel Multi-Stage Biomass Thermal Fractionation Process for Fuels and Chemicals

Abstract:

In this work we present our research into a novel, multi-stage thermochemical biomass fractionation strategy. We demonstrate the ability of this process to deliver multiple streams of bio-oil which are better suited for further downstream catalytic upgrading than single-stage fast pyrolysis bio-oil. Process optimization techniques and feedstock development possibilities are also discussed.

443979 GHS: OSHA's Final Deadline -- Last Call for Employer Compliance

Abstract:

OSHA's GHS transition period is set to end in 2016. By June 1 of this year, OSHA expects millions of HazCom-covered employers to be in full compliance with its fourth and final GHS deadline. This means having an updated HazCom Program in place that accounts for the GHS changes OSHA implemented when it adopted the United Nations' Globally Harmonized System in 2012.

This might sound straightforward, but adoption delays by chemical manufacturers and suppliers, as well as recent OSHA memorandums and selective enforcement rules, compromised earlier deadlines, and for many, created confusion in the chemical supply chain. Additionally, enforcement statistics have shown that just complying with the basic tenets of OSHA's HazCom Standard remains challenging for many employers.

In fact, the standard continues to rank #2 on OSHA's most frequently cited standard list. Compliance is important for avoiding fines and maintaining a positive corporate reputation, but it's probably most important for protecting employees from the increased injury and illness risks they face as a result of a compromised employer HazCom Program.

In this session, find out what it means to be in compliance with OSHA's June 1, 2016 employer deadline, what OSHA's general compliance expectations are, the common compliance challenges impacting EHS professionals and how to overcome them, and the actionable steps you can take to ensure compliance with OSHA's fourth and final GHS deadline and beyond.

444902 Rapid Assessment Tool for Process Safety Management (PSM) Systems

Abstract:

Senior management at facilities with significant process hazards have two key responsibilities related to process safety management (PSM). These responsibilities are:

(1) Ensure that the design of the management systems for process safety is adequate;

(2) Ensure that the management systems are implemented as designed.

As part of the second responsibility, full compliance audits are one method of evaluating PSM systems, but this type of audit can be resource-intensive. Besides, companies are reluctant to carry out compliance audits because of the cost and planning required for their execution. This presents a problem as areas of relatively high process risk may exist which are not effectively managed.

This presentation describes a simple assessment tool designed to rapidly and effectively identify areas where a process safety management system is doing well, and areas where opportunities for improvement in the system exist. Using the tool is not a substitute for a full assessment, but a way to ensure that process safety execution is going in the right direction, and addresses the areas with the most need for improvement first. The tool focuses on providing support to senior leadership.

The tool consists of 10 questions which are posed by an assessor who is an experienced process safety specialist, to be answered by high level site management and their representatives. The questions are based on generally accepted knowledge of the key drivers of a successful and self-sustaining PSM program. The answers provided to the questions, their analysis, discussion and the practical recommendations for improvement provided by the assessor will be useful to site management in designing and implementing a more effective PSM system.

442698 Safer and Smarter Sulfiding of Hydrotreating Catalysts

Abstract:

Catalyst sulfiding (aka "presulfiding") is a critical step to ensure that hydrotreating catalysts perform as designed after startup. In order to optimize the sulfiding process, Reactor Resources (RxR) has developed a new DMDS/TBPS Injection System that allows the refiner to continuously monitor sulfur injection rates, H2S levels, and Hydrogen Purity in real time during sulfiding operations. These mobile units, called SmartSkid™ Injection Systems, are equipped with wireless telemetry that is streamed in real-time via the cloud. With "live telemetry", engineers and operations personnel can monitor pump operations and recycle gas concetrations from any computer or smart phone.

This allows customers to constantly monitor parameters such as:

- Flow rate of the sulfur spiking agent
- Total quantity of chemical injected
- Pressure
- H2S Concentration
- Hydrogen Purity
- Fluid Temperature
- Valve Position
- Pump Speed

By utilizing an Online H2S Analyzer System to continuously measure the H2S concentration of the recycle gas, the hazardous practice of operations personnel taking draeger tube measurements while equipped with breathing air (SCBA) is eliminated. And with real time H2S data, sulfur injection rates can be controlled more precisely, avoiding over-injection of sulfur chemicals that can lead to compressor issues and the need to purge sour recycle gas. The online Hydrogen Purity Analyzer measures the hydrogen concentration of the treat gas in real time, which also helps to avoid compressor trips and excessive purging that can lead to excess sulfur emissions.

444112 Case Study of Accidental Chlorine Release in the Urban Area Using CFD and Acceleration of the Simulation

Abstract:

Computational Fluid Dynamics is widely used to simulate accidental hazardous chemical releases. In the urban area where dispersion obstacles are large and abundant, CFD would be the best choice to simulate the toxic chemical dispersion. In this paper, one industrial complex which is adjacent to residential area was chosen for the simulation and tested for 10 cases of the wind condition. One worst case was selected and solved for 1-hour after the chlorine release begins. And then for the vulnerable points in the simulation volume, toxic load and probability of death was calculated. After that, using Courant-Friedrich-Levy number, time stepping scheme of the implicit Euler method was modified to increase timte-linearly to make simulation faster but to keep robustness. With combination of linear CFL number and parallel solvers, CPU time of the CFD simulation became one-forth of the base case.

Experimental determination of physicochemical characteristics of new environmental gasoline, ethanol and isopropanol blends

"Seeking a clean and healthy environment"

Tarek M. Aboul-Fotouh and Mohamed S. Hussein

Mining & Petroleum Engineering Department, Faculty of Engineering, Al-Azhar University, Cairo, Egypt.

E-mail: tarekfetouh@yahoo.com

Abstract

The increasing number of automotive vehicles around the world is responsible for the emission of great amounts of pollutant gases, which can cause serious damage to environment and to human health. Among the fuels used in light vehicles the gasoline, which is a blend of "n-paraffin, naphthenes, aromatics, n-butane, and some additives". Some of these groups (aromatics 45-50%) are used to increase the octane number, which is considered as a key factor in the fuel quality. However, the aromatic compounds in this fuel contain benzene rings; A Carcinogenic compound which is proved to cause a lot of serious damage to the human health may cause lung cancer. Also the emissions contain amounts of carbon monoxide (CO), nitrogen oxides (NO_x) and sulfur compounds which may raise serious environmental problems. So new regulations (Euro-6) were made to determine the safe ranges of each component in the blend. After the new regulations, the percent of aromatics must not exceed a maximum limit of 35% to keep the benzene rings percent in the blend equal to or less than 1%. In this paper we investigate the effect of adding new compounds to increase the octane number within the safe ranges. This paper presents the high octane number environmental gasoline which is a blend of some products: "Naphthas of hydrotreated and hydrocracking, Reformate, Isomerate, and Ethanol and Isopropyl alcohol as oxygenated compounds". The paper also illustrates the exact percent of each component in the blend, the research octane number and the motor octane number of the blend, and all the physicochemical characteristics of the blend, reaching the optimum three samples which nearly approach the standard A92 and A95 gasoline blends. Hopefully, this paper will attract the attention and promote seeking a clean air and safe environment for the next generations.

Keywords: Ethanol, Isopropyl Alcohol, Euro-6, Environmental Gasoline, Physicochemical Characteristics of Gasoline Blends.

1. Introduction

The European Union has given a major importance to climate and energy issues, not only for environmental reasons, but also with a geostrategic aim, in order to reduce excessive dependence of imports of energy resources [1]. This can be seen in the 20/20/20 targets to reduce climate impacts and energy consumption. The final target is to achieve a low carbon economy that promotes green growth and to strength the Europe's competitiveness. Specifically, these targets aim to achieve by 2020 a 20% reduction in EU greenhouse gas (GHG) emissions from 1990 levels; raising the share of EU energy consumption produced from renewable resources to 20% and a 20% improvement in the EU's energy efficiency.

Many trials are made to enhance the properties of gasoline in order to produce an Eco-friendly gasoline such as:

EL-Bassiouny et al. [2,3] investigated the Maximize of the Production of Environmental, Clean and High Octane Number Gasoline-Ethanol Blends by using Refinery Products. Moreover, Straight run naphtha, isomerate, reformate, Coker naphtha and hydrocracked naphtha which produced from crude distillation unit, isomerization process, catalytic reforming and conversion processes respectively are blended with an oxygenated compound (ethanol) to produce environmental gasolines A98 and A95 which satisfy all specifications of Euro-6 regulations. The new blends are friendly environmental and contains the less amount of benzene content. It is the best solution for the long run. Moreover, EL-Bassiouny et al. [4] studied the upgrading of the commercial gasoline A80 by using ethanol and refinery products by using blends of gasoline A80 and Refinery products from upgrading and conversion units and thus the Environmental gasoline is produced to approach to the standard European regulations of A92. In other words, three samples of reformate, isomerate, Coker naphtha and hydrocracked naphtha are blended separately with gasoline A80. In addition, the three samples of oxygenated compounds (Ethanol) and gasoline A80 are blended to produce the environmental gasoline. Furthermore, Altun et al. [5] studied the effect of 10 vol. % of ethanol and 10 vol. % of methanol blending in unleaded gasoline on engine performance and exhaust emission. Results indicated that M10 and E10 blended fuels demonstrated the best result in the exhaust emission. The HC emission of M10 and E10 are reduced by 13% and 15% and the CO emissions by 10.6% and 9.8%, respectively. Increased CO_2 emission for M10 and E10 compared with unleaded gasoline was observed. The ethanol and methanol addition to unleaded gasoline demonstrated an increase of BSFC (brake specific fuel consumption) and a decrease of brake thermal efficiency in comparison to unleaded gasoline. In other paper, Kheiralla et al.[6] investigated the experimental determination of fuel properties of gasoline-ethanol blends as bio-fuel for SI engines. Using standard laboratory methods, fuel properties of selected blends such density, API gravity, kinematic viscosity, cloud point, flash and fire point, heat value, ASTM distillation and Octane number were studied and compared to gasoline fuel. In addition, Abdel-Rahman and Osman [7] recently tested 10, 20, 30 and 40 vol. % respectively ethanol of blended fuels in a variable-compression-ratio engine. They found that the increase of ethanol content increases the octane number but decreases the heating value. The 10 vol. % addition of ethanol had the most obvious effect on increasing the octane number. Under various compression ratios of the engine, the optimum blend rate was found to be 10 vol. % ethanol with 90 vol. % gasoline. Besides, Hakan Bayraktar [8] investigated the effects of ethanol addition to gasoline on an SI engine performance and exhaust emissions experimentally and theoretically. Experimental applications have been carried out with the blends containing 1.5, 3, 4.5, 6, 7.5, 9, 10.5 and 12 vol. % ethanol. Experimental results have shown that among the various blends, the blend of 7.5 vol. % ethanol was the most suitable one from the engine performance and CO emissions. However, theoretical comparisons have shown that the blend containing 16.5 vol. % ethanol was the most suited blend for SI engine.

Rodríguez-Anton et al.[9] Evaluated the influence of the simultaneous addition of ethanol and ETBE on some physical properties of engine gasoline. The main conclusion is that the simultaneous addition of ETBE and ethanol changes the RVP, ASTM distillation curve, and the density in a way that can affect engine operation and the mandatory EN 228 and ASTM D4814 standards. Some opposite properties of both oxygenates could help to increase the renewable energy content without preventing compliance with these regulations .Furthermore, Muzíková [10] studied the effect of ethanol up to 10 vol. %, ETBE up to 10 vol. % and hydrocarbon composition over volatility and ASTM distillation. In addition, Petre [11] used two classical gasolines with different compositions and properties, and with proportions ranging from 2 to 10 vol. % of ethanol and other alcohols, presents experimental results regarding the effect of the blend with alcohol on the RVP, ASTM distillation curves and evaporated percent at 70 °C. In another paper [12] by the same author and used two classical gasolines with different compositions and properties, and with proportions ranging from 4 to 15 vol. % of ETBE and other ethers, she presents experimental results regarding the effect of the blend with alcohol on the RVP, ASTM distillation curves and vapor lock index while Pumphrey [13] offered a simple method to successfully predict vapor pressures of gasoline–alcohol mixtures and validate them with blends of gasoline with different oxygenates (0–100) vol. %.

Soheil et al. [14] studied the effect of oxygenate additives into gasoline for the improvement of physico-chemical properties of blends. Methyl Tertiary Butyl Ether (MTBE), Methanol, Tertiary Butyl Alcohol (TBA), and Tertiary Amyl Alcohol (TAA) blend into unleaded gasoline with various blended rates of 2.5, 5, 7.5, 10, 15, and 20 vol. % respectively. Physico-chemical properties of blends are analyzed by the standard American Society of Testing and Materials (ASTM) methods. On the other hand, Khamis and Palichev [15] investigated the production of stock gasoline with ultra-low sulphur content up to 10 ppm (Euro-5 Standard) by blending of different gasoline streams produced in the Lukoil Nentochim Bourgas (LNHB) refinery units as well as on the study on the efficiency of ferrocene antiknock additives. Some recipes for the production of stock gasoline A92, A95, and A98 commercial grade on the basis of component streams produced in LNHB refinery units and satisfying all specifications of the European regulations were elaborated. Thus, the gasoline blending provides a great potential benefit to the refinery in view of minimizing operating costs and product quality improvement.

The latest standard regulations in the world set a maximum limit for the aromatics compounds in the fuel due to its serious health effect, the octane number must be raised by other means, and thus two approaches have been investigated in this paper. The first one is optimizing the use of components like (isomerate, reformate and hydrocracked naphtha) restricted by the available local supply using many different volume combinations and then the aromatic content is estimated in each sample till obtaining the optimum blending volumes percent. In addition, the second approach is blending high octane oxygenates, Ethanol and Isopropanol, which have been added with different volumes percent to the base gasoline sample. Clearly, the results represent the effect of oxygenated compounds on physicochemical properties of gasoline blends; a remarkable enhancement in ASTM distillation curves of gasoline blends.

2. Materials and methods

The purpose of this study is to present the isopropanol-gasoline-blend fuel and comparing its physicochemical characteristics such as density, RON, MON, RVP, ASTM distillation, PIONA and benzene content with the most widely known ethanol-gasoline-blend fuel.

2.1. Gasoline blends

This research is carried out on the following blend stocks:

1. Hydrotreated Naphtha.
2. Reformate.
3. Isomerate.
4. Hydrocracked Naphtha.
5. Ethanol.
6. Isopropanol.

2.2. Standard methods

Fuel properties of tested blends are determined in accordance with American Standard for Testing and Materials (ASTM) procedures for petroleum products. Comprehensive analyses are carried out to document fuel properties of the tested blends. Each fuel sample is evaluated to determine the following physico-chemical characteristics such as the density, RVP, ASTM distillation, RON, MON, PIONA and benzene content. Figs 1-6 list various apparatus employed in the determination of these fuel properties. The density of each tested sample is measured by density meter apparatus (ASTM 4052) [16]. The distillation curve is determined by ASTM distillation device in accordance with ASTM D86 [17]. The Reid vapor pressure is determined by Reid vapor pressure device in accordance with ASTM D 5191[18]. The blends Octane rating is determined by a Cooperative Fuels Research (CFR) Engine and octane meter apparatus for research octane number (ASTM D2699)[19] and motor octane number (ASTM D 2700) [20]. The Hydrocarbon types are obtained by Gas chromatography in accordance with ASTM D 6839 [21].

Fig.1 Experimental Blended Samples.

Fig. 2 Cooperative Fuels Research (CFR)Engine.

Fig. 3 Gas Chromatography Apparatus.

Fig.4 Distillation Apparatus.

Fig. 5 Reid Vapor Pressure Apparatus.

Fig. 6 Density Meter Apparatus.

3. Results and discussion

In this section, the main results obtained from the tests (RVP, RON, MON, ASTM distillation curve, density and composition) are shown and discussed.

Table 1 demonstrates PIONA composition for determining the amount of Parrafins, Olefins, Aromatics, isoparaffins, Naphthenes and benzene content of different sources of blend stocks of categories A and B. Blend stocks are completely blended in order to produce different refinery gasoline - blend samples which have different compositions of Hydrotreated Naphtha, Reformate, Isomerate and Hydrocracked Naphtha as shown in Table 2. Tables 3 and 4 show PIONA compositions of fifteen gasoline-blend samples in categories A and B. The blend of sample number seven has been selected according to the isoparaffins, aromatics and benzene contents; approach to specifications of Euro-6 (categories A and B) [22].

Table1 The Blend Stocks Composition(PIONA) by Using Gas Chromatography.

Blend Stocks	Category A						Category B					
	P	I	O	N	A	Benzene	P	I	O	N	A	Benzene
HT Naphtha	35.9	31.6	0.1	22.6	9.8	1	33.7	33.9	0	22.7	9.7	1
Isomerate	12.5	73.2	0	14.3	0	0	12.5	73.3	0	14.3	0	0
Reformates	9.9	17.9	1.3	2.6	68.3	1.3	7.3	16.2	1.7	0.9	73.9	2.8
HC Naphtha	24.8	42.8	.05	29.95	2.4	1.2	23.4	37.8	0	30.7	8.1	1.4

Table 2 Gasoline-Blend Samples with Different Combinations.

Blend Stocks (vol.%)	Gasoline-Blend Samples														
	1	2	3	4	5	6	7	8	9	10	11	12	13	14	15
HT Naphtha	15	10	13	17	10	10	10	13	9	10	10	9	9	13	9
Isomerate	15	12	13	15	10	15	12	10	10	10	10	10	9	12	9
Reformate	40	47	40	35	45	40	42	42	42	45	42	42	45	35	42
HC Naphtha	30	31	34	33	35	35	36	35	39	35	38	39	37	40	40
SUM	100	100	100	100	100	100	100	100	100	100	100	100	100	100	100

Table 3 The Composition of the Tested Samples by Using Gas Chromatography.

Category (A)	Test samples														
	1	2	3	4	5	6	7	8	9	10	11	12	13	14	15
P	18.66	17.74	18.68	19.63	17.98	18.11	16.68	18.76	18.31	17.98	18.42	18.31	17.99	19.55	18.43
I	35.72	33.63	35.34	36.74	33.52	36.28	34.87	33.93	34.37	33.52	34.26	34.37	33.32	36.28	34.07
O	.55	0.64	.55	.49	.61	0.55	0.57	0.58	0.57	0.61	0.58	0.57	0.61	0.49	0.58
N	15.56	14.48	16.02	16.78	15.34	15.93	14.37	15.94	16.24	15.34	16.16	16.24	15.57	17.54	16.39
A	29.51	33.83	29.41	26.36	32.56	29.14	33.53	30.8	30.5	32.56	30.58	30.5	32.51	26.14	30.53
Benzene	1	1.1	1.1	1	1.1	1	1.1	1.1	1.1	1.1	1.1	1.1	1.1	1.1	1.1

Figure 7 illustrates the Relationship between Research octane number and the percentages of Ethanol and Isopropanol in the gasoline blend samples (Category A & B) .It indicates 94.5 and 91.5 RONs at 20 and 15 volume percents of ethanol respectively for category A. In addition, it represents 91.8 RON at 20 volume percent of isopropanol for the same category. On the other hand, it shows 95 and 92 RONs at 20 and 15 volume percents of ethanol respectively for category B while it indicates 92 RON at 20 volume percent of isopropanol. Figures 8 and 9 illustrates RON and MON of ethanol and isopropanol in gasoline-blend samples for categories A and B.

Table 4 The Composition of the Tested Samples by Using Gas Chromatography.

Category (B)	Test samples														
	1	2	3	4	5	6	7	8	9	10	11	12	13	14	15
P	16.87	15.56	16.88	17.88	16.1	16.35	16.8	16.89	16.48	16.1	16.58	16.48	16.1	17.8	16.58
I	33.89	31.51	33.26	34.89	31.23	34.08	32.2	31.76	31.92	31.23	31.88	31.92	30.92	33.98	31.56
O	0.68	0.8	0.68	0.6	0.77	0.68	0.8	0.71	0.71	0.77	0.71	0.71	0.77	0.6	0.71
N	15.12	13.93	15.61	16.45	14.85	15.52	14.5	15.5	15.82	14.85	15.74	15.82	15.09	17.26	15.95
A	33.45	38.21	33.58	30.19	37.06	33.37	35.7	35.13	35.07	37.06	35.09	35.07	37.13	30.37	35.15
Benzene	1.7	1.9	1.7	1.6	1.9	1.7	1.6	1.8	1.8	1.9	1.8	1.8	1.9	1.7	1.8

Table 5 Physico-chemical Characteristics of Optimum Gasoline Blend Samples.

Physicochemical Characteristics	Standard Regulations		Optimum Results – Category A		
	A92	A95	Ethanol (15 vol.%) A92	Ethanol (20 vol.%) A95	Isopropanol (20 vol.%) A92
Density@15°C, g/cc	0.720 – 0.775	0.720 – 0.775	0.7541	0.7561	0.7542
RON	mini. 92	95	91.5	94.5	91.8
MON	mini. 82	85	86.2	89.2	87.9
Benzene, vol.%	max. 1	max. 1	0.9	0.8	0.8
Aromatic, vol.%	max. 35	max. 35	29.3	29.7	28.5
Distillation Characteristics:					
%evaporated @100°C, vol.%	46 – 71	46 – 71	57	65	63
%evaporated @150°C,vol%	min. 75	min. 75	90	90	90
FBP, °C	max. 210	max. 210	196.5	196	196
Reid Vapor pressure, kPa	45 -60	45 - 60	49	49	44

Figure 10 shows 29.3 and 29.7 volume percents of aromatic contents at 15 and 20 volume percents of ethanol in category A and benzene contents are 0.9 and 0.8 respectively. Hence, these samples are chosen as optimum gasoline blends according to Euro-6 (Table 5). However, the previous contents are 28.5 and 0.8 volume percents of isopropanol blend sample respectively and thus it has been chosen as an optimum gasoline blend too (Table 5). All results in Figures 11 and 12 are identically for the ranges of Euro-6 regulations regarding to distillation at 100 and 150 °C and the final boiling point (Table 5). All values of densities and Reid vapor pressures are represented in Figure 13 for blended samples of category A. In addition, all results of optimum samples are in the ranges of Euro-6 regulations.

4. Conclusions

Ethanol and isopropanol in gasoline blends can be used as alternative fuels that are high octane number environmental gasolines (95 and 92 RONs). The optimum samples obey to the standard European regulations (Euro-6) and thus this work can contribute to maximize the quality and quantity of environmental gasolines. Finally, environmental gasolines provide a great potential benefit to the refinery in view of minimizing operating costs, product quality improvement, safe and healthy living environments.

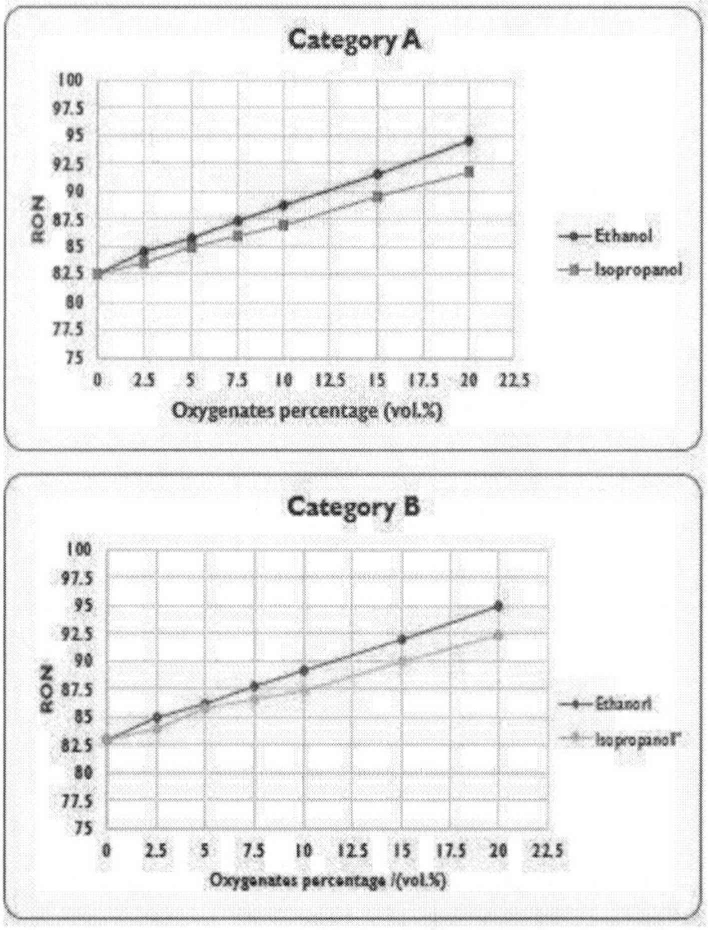

Fig. 7 Relationship between Research Octane Number and the Percentage of Ethanol and Isopropanol in the Gasoline Blend Samples (Category A & B).

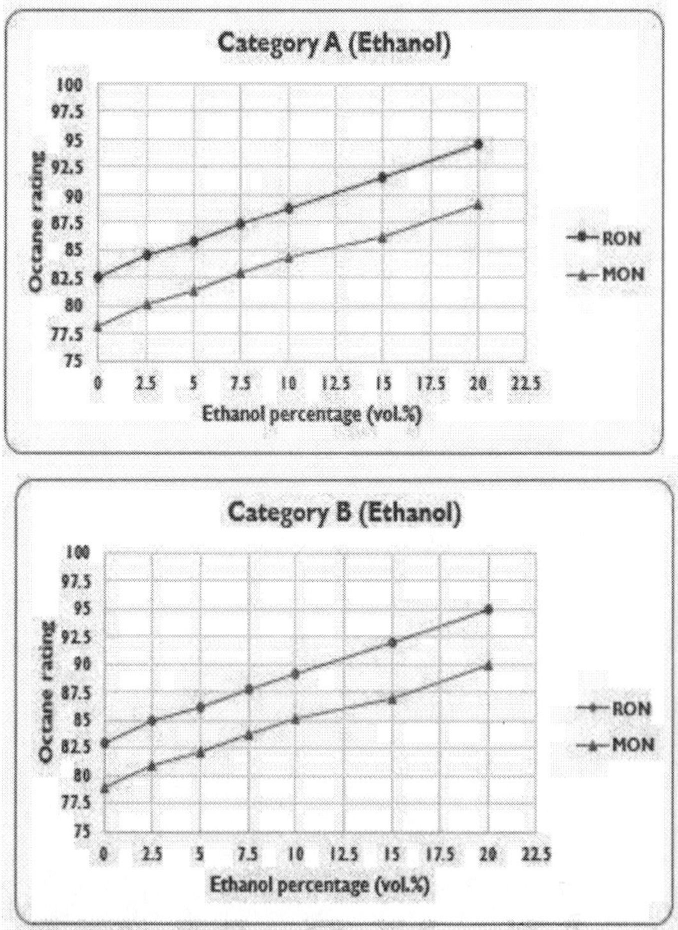

Fig. 8 Relationship between Octane Numbers and Gasoline - Ethanol Blends (Category A & B).

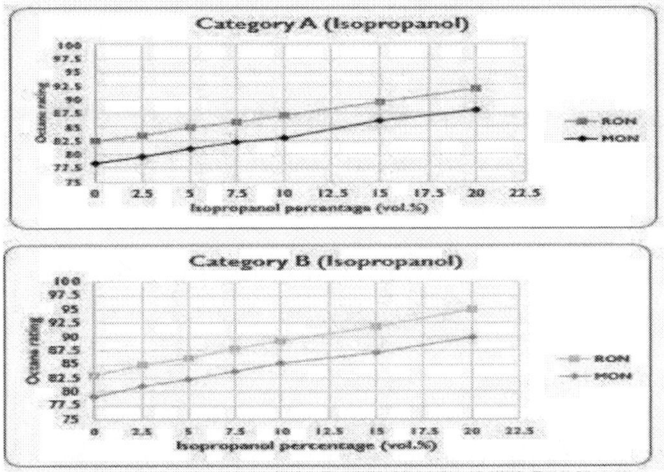

Fig. 9 Relationship between Octane Numbers and Gasoline - Isopropanol Blends (Category A & B).

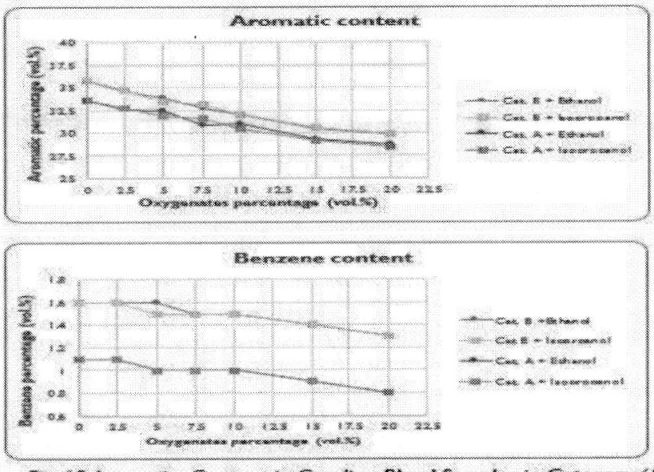

Fig. 10 Aromatic Content in Gasoline-Blend Samples in Category (A & B).

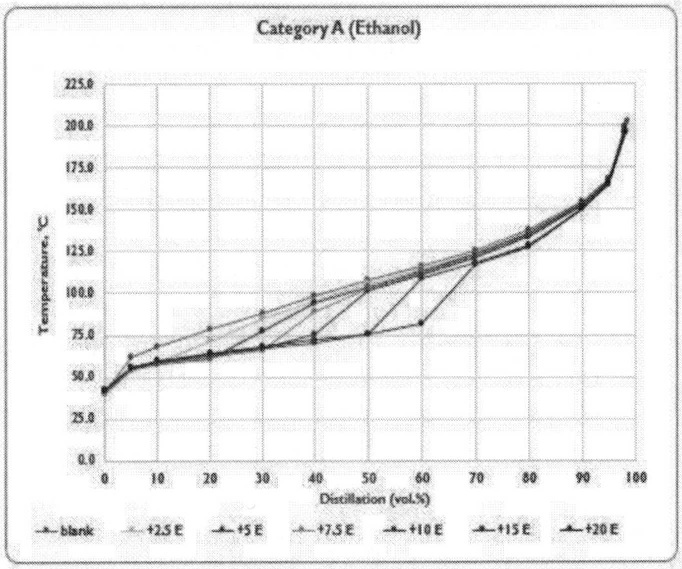

Fig. 11 ASTM Distillation Curves for Gasoline - Ethanol Blend Samples (Category A).

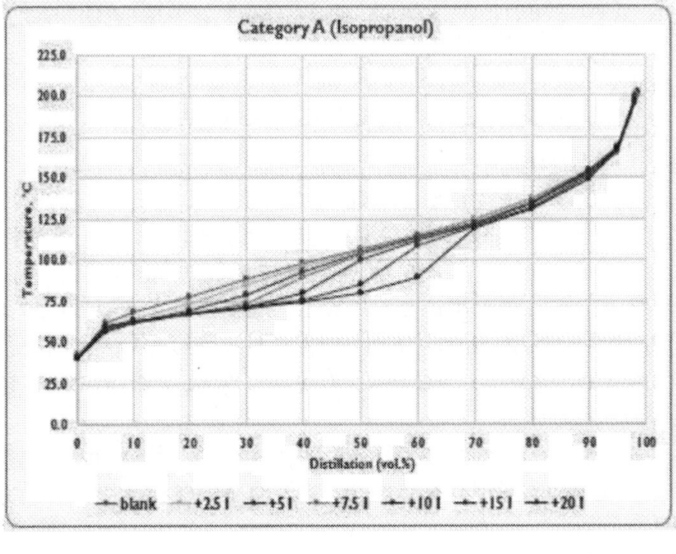

Fig. 12 ASTM Distillation Curves for Gasoline - Isopropanol Blend Samples (Category A).

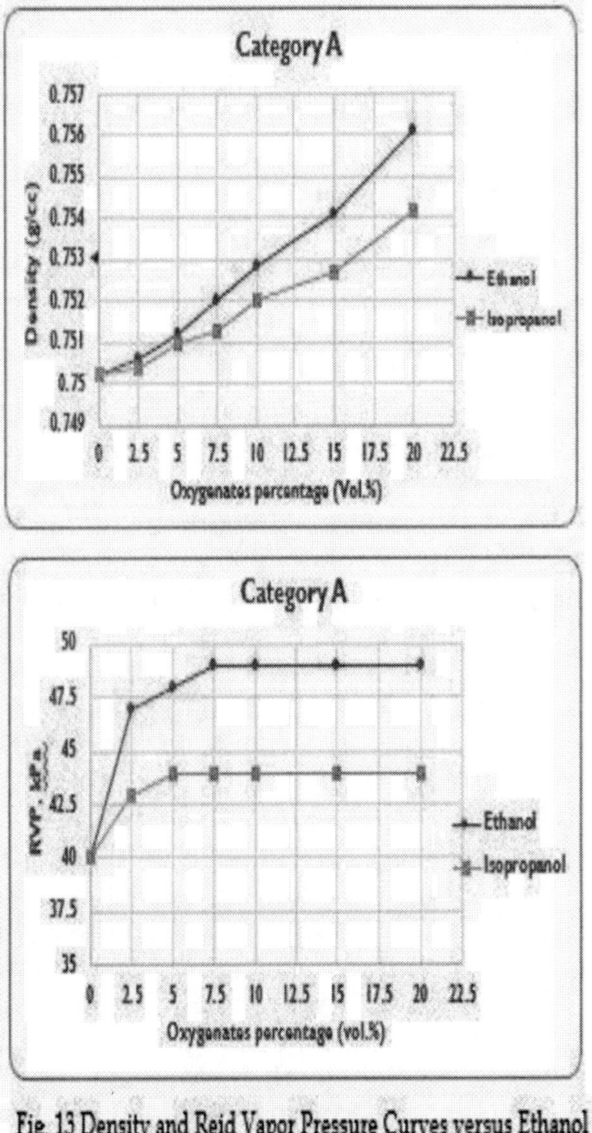

Fig. 13 Density and Reid Vapor Pressure Curves versus Ethanol and Isopropanol Percentages in Gasoline Blends.

5. References

1) ENT Environment and Management C/Sant Joan, 39, 1r. Vilanova i la Geltrú. 08800 Barcelona, Climate and energy 20/20/20 targets of the European Union and the failure of energy efficiency objectives, 2012. http://ec.europa.eu/europe2020/index_en.htm.
2) Abdel-Monem A. El-Bassiouny, Tarek M. Aboul-Fotouh and Tamer M. M. Abdellatief, "Maximize the Production of Environmental, Clean and High Octane Number Gasoline-Ethanol Blends by using Refinery Products ", International Journal of Scientific & Engineering Research, Volume 6, Issue 7, 1792-1803, USA, July-2015.
3) Tamer M. M. Abdellatief, Abdel-Monem A. El-Bassiouny and Tarek M. Aboul-Fotouh, " An environmental Gasoline (Enhancing the properties of the gasoline through modified blending operations) ", Lab Lambert Academic Publishing,1-121, Germany, November- 2015.
4) Abdel-Monem A. El-Bassiouny, Tarek M. Aboul-Fotouh and Tamer M. M. Abdellatief, "Upgrading the Commercial Gasoline A80 by Using Ethanol and Refinery Products", International Journal of Scientific & Engineering Research, Volume 6, Issue 8, 405-417, USA, August-2015.
5) S. Altun, H. Oztop, C. Oner, Exhaust and Y. Varol, "Emissions of methanol and ethanol-unleaded gasoline blends in a spark ignition engine", Thermal science, 17, 1, pp. 291-297, 2013.
6) A. F. Kheiralla, M.M. El-Awad, M.Y. Hassan, M. A. Hussen, and I. Hind, "Experimental Determination of Fuel Properties of Ethanol/Gasoline Blends as Bio-fuel for SI engines", International Conference on Mechanical, Automobile and Robotics Engineering (ICMAR'2012)Penang. Malaysia; 244- 249, 2012.
7) A. A. Abdel-Rahman and M. M. Osman," Experimental investigation on varying the compression ratio of SI engine working under different ethanol–gasoline fuel blends" , International Journal of Energy Research 21, 31–40, 2006.
8) H. Bayraktar, "Experimental and theoretical investigation of using Gasoline–ethanol blends in spark-ignition engines", Renewable Energy 30, 1747-1733, 2005.
9) L.M. Rodríguez-Anton, F. Gutíerrez-Martín, C. Martinez-Arevalo, "Experimental determination of some physical properties of gasoline,ethanol and ETBE ternary blends" , fuel, 2015. http://dx.doi.org/10.1016/j.fuel.2015.04.040.
10) Muzíková Z, Pospíšil M, Šebor G. "Volatility and phase stability of petrol blends with ethanol". Fuel; 88:1351–6, 2009.
11) Petre M, Rosca P, Dragomir RE, Mihai O. "Bio alcohols – compounds for reformulated gasolines. The effect of alcohols on volatility properties of gasolines". Rev Chim-Bucharest; 61:706–11, 2010.
12) Petre M, Rosca P, Dragomir RE. "The effect of bio-ethers on the volatility properties of oxygenated gasoline of oxygenated gasoline". Rev Chim-Bucharest; 62:567–74, 2011.
13) Pumphrey JA, Brand JI, Scheller WA. "Vapor pressure measurements and predictions for alcohol–gasoline blends". Fuel; 79:1405–11, 2000.
14) B. S. Soheil, S. S. Morteza, and O. Fathollah, " Effect of Oxygenates Blending with Gasoline to Improve Fuel Properties ", chinese journal of mechanical engineering, Vo l. 25, No. 4, 2012.

15) F. Khamis and T. Palichev, "Production of Ultra-low Sulphur Gasoline and Assement of the Efficiency of Ferocene Antiknock Additives",International Journal of Engineering and Applied Science, Vol.1 , No. 1 ,November 2012.
16) American Society for Testing and Materials. D4052 Standard Test Method for Density, Relative Density, and API Gravity of Liquids by Digital Density Meter, 2011.
17) American Society for Testing and Materials. D86-04 Standard Test Method for Distillation of Petroleum Products at Atmospheric Pressure, 2011.
18) American Society for Testing and Materials. D5191 Standard Test Method for Vapor Pressure of Petroleum Products. (Reid Method), 2011.
19) American Society for Testing and Materials. D2699 Standard Test Method for Research Octane Number of Spark-Ignition Engine Fuel, 2011.
20) American Society for Testing and Materials. D2700 Standard Test Method for Motor Octane Number of Spark-Ignition Engine Fuel, 2011.
21) American Society for Testing and Materials. D6839 Standard Test Method for Gas Chromatography of Petroleum Products, 2011.
22) European Standard EN 228, Automotive fuels - Unleaded petrol - Requirements and test methods, 2008.

444407 Multi-Phase CFD Simulation of a Downdraft Biomass Gasifier

Abstract:

A reliable, affordable clean energy supply is of primary importance for the economy and the environment and will prove to be pivotal in the 21st century. Biomass energy which is now accepted as a propitious alternative includes a widely available, sustainable and as a neutral CO2 resource, suited for modern applications for electricity generation and production of fuels and chemicals.

An advanced multi-dimensional mathematical model has been developed to understand the configuration and operation of a fixed bed biomass gasifier considering drying, pyrolysis, combustion and gasification reactions. The Computational Fluid Dynamics (CFD) simulation is based on the down draft biomass gasifier design operated by the Energy Research and Development center at Missouri University and Science and Technology in Rolla, Missouri. The multi-phase CFD model has been developed and verified with results from the experimental work. The flow model assumes an Eulerian –Eulerian formulation of the multi-phase system and uses kinetic theory of granular flow as the basis for the approximation to the turbulence flow in the solid phases with the standard κ-ε turbulence model used for the gas phase.

Results illustrate the influence the air ratio has on gasification performance in terms of the production of syn gas and the amount of char produced. Predictions of gas temperature distribution through the gasifier and product gas composition are also compared to experimental results. Comparisons between the simulation and experimental results have been used to evaluate the validity of the CFD based gasification model.

444863 Preliminary Evaluation of High-Pressure Circulating Fluidized Bed Biomass to Fuels Technology

Abstract:

Pyrolysis of second generation biomass, such as crop residues or non-food crops, seems to be a very promising option for generating clean, renewable, drop-in biofuels. Presently, there are several pyrolysis technologies that must be evaluated to determine the most cost effective technology to use for commercialization as the United States energy catalog lacks an effective way to convert biomass into useful liquids for fuels and other uses. Southwest Research Institute® (SwRI®) employees have devised a way to combine circulating fluidized bed reactor conversion with a combination of catalysts and conditions to efficiently convert general biomass into a broad-boiling point range product ready for refining into transportation fuels.

The University of Texas at San Antonio (UTSA) and SwRI through a cooperative program CONNECT are currently working on a project to develop and demonstrate a novel biomass conversion technology (high-pressure circulating fluidized bed hydropyrolysis) to enable the goal of making biofuels cost-competitive with corn-based ethanol and petroleum based fuels. This project aims to improve the performance and test the commercial viability of this new technology considering uncertainties and risks and taking a holistic cradle-to-grave approach to test the economic viability of producing biofuels and other biobased products (byproducts).

This presentation/paper analyzes the experimental results obtained from the preliminary bench scale experiments completed to validate SwRI's initial design. Experiments to evaluate SwRI's proposed technology at bench scale are addressed by performing a set of experiments in SwRI's high-pressure, high-temperature, Parr reactor system. One type of feedstock (switchgrass) is tested at supercritical and subcritical pressures, with and without sodium carbonate, and at various catalyst blend ratios. A mass balance is completed for each run including solid, liquid, and gas yields and analysis of products and feedstock.

444751 Catalytic Production of BTX from LPG

Abstract:

A systematic study of the comparative performances of 1 wt. % and 3 wt. % Ga, 1 wt. % and 3 wt. % Zn, 3 wt. % Mo, and 3 wt. % Re impregnated ZSM-5 catalysts for propane aromatization is presented. It was found that methane, ethene, ethane, benzene, toluene, and xylene are the major products on all of these catalysts at typical propane aromatization temperature (~550 °C). 3% Zn/ZSM-5 catalyst shows the highest propane conversion among all the tested catalysts with approximately 56% total selectivity toward benzene, toluene, and xylene (i.e., BTX) but lower selectivity to methane, ethene, and ethane (i.e., light hydrocarbons). In sharp contrast, 3% Mo/ZSM-5 catalyst shows the lowest activity with lower BTX selectivity but as high as 52% total selectivity toward methane, ethene, and ethane. Further, calculations indicate that 1% Zn/ZSM-5 catalyst displays the highest H_2 and BTX formation rates (mmol/s/mol–metal), followed by 1% Ga, 3% Re, 3% Zn, and 3% Ga impregnated ZSM-5 catalysts, while 3% Mo/ZSM-5 catalyst shows remarkably lower H_2 and BTX formation rates. This suggests that the addition of 1% Zn into ZSM-5 via impregnation significantly improves the dehydrogenation and subsequent chain growth reaction as well as cyclization, thereby favoring propane aromatization. However, compared to other promoters (i.e., Zn, Ga, and Re), the promotional effect of Mo over ZSM-5 is insignificant for propane aromatization, thus resulting in substantial formation of methane, ethene, and ethane.

444602 Next Generation Phenol Technology

Abstract:

The Next Generation Phenol technology offers phenol producers with growth prospects and/or producers pursuing vertical integration of their refining & petrochemical assets via phenol to bis-phenol A to polycarbonate production a lower OPEX, CAPEX, and smaller footprint processing option. Economic advantages include a reduction of $21 per ton of phenol in net cost of production (NCOP), and at least a $19M increase in net present value (NPV) relative to current state-of-the art process alternatives. This is achieved by: 1) a reduction in utilities of 10% using novel phenol recovery fractionation technology that integrates the Cumene-AMS and Crude-Acetone columns, 2) an improved cumene oxidation/ CHP decomposition design that reduces waste water by one half and produces higher phenol yield and reduced CAPEX, and 3) offers a further improvement in feedstock efficiency when combined with the Q-Max cumene alkylation technology using a new QZ-2500H high selectivity alkylation catalyst.

444080 Production of Benzene, Toluene, and Xylenes from Natural Gas Via Methanol: A Comprehensive Process Synthesis and Global Optimization Approach

Abstract:

Recently discovered sources of shale gas, along with the technological improvements within the shale gas industry, have greatly increased the supply of natural gas in the United States. However, several challenges have emerged due to these developments, a critical one being the proper utilization of this resource[1] into more valuable and highly demanded products via routes that are both economical and profitable. Thus, we have developed a process synthesis framework capable of comparing several commercial and promising technologies for the production of aromatics from natural gas via methanol.

Between 2005 and 2008, the global demand for benzene, para-xylene, ortho-xylene, and meta-xylene was approximately 40, 26, 6, and 0.4 million metric tons per year, respectively.[2] These aromatic commodities have several applications within industry. Benzene is used in styrene, phenol, nylon, and aniline production. Ortho-xylene is used for the production of phthalic anhydride, while meta-xylene is converted into isophthalic acid. Para-xylene, the most valuable xylene isomer, is converted into terephthalic acid and dimethyl terephthalate, which are ultimately used to produce polyethylene terephthalate (PET) fibers, resins and films. Thus, opportunities exist for widespread penetration of natural-gas based chemical refineries, albeit with challenges associated with which types of chemicals to produce and which technologies to invest in.

In this work, we propose a process synthesis superstructure that contains several direct and indirect commercialized natural gas conversion technologies, including autothermal reforming and steam reforming, among others.[3,4] To address the economic barriers often associated with the utilization of alternative feedstocks, we develop a novel branch-and-bound global optimization framework that is capable of determining the optimal technologies to produce aromatics from natural gas at the highest profit.[3-6] Several novel, commercial, and/or competing technologies are modeled within the framework, including methanol-to-aromatics, toluene alkylation with methanol, selective toluene disproportionation, toluene disproportionation and transalkylation with heavy aromatics, para-xylene separation via adsorptive separation or crystallization, isomerization of xylenes, and dehydrocyclodimerization of liquefied petroleum gas, among others.[6] This approach provides an adequate baseline for comparing competing technologies and identifying bottlenecks for natural gas conversion based technologies. Several case studies are presented that explore the effect of refinery capacity and product output. The key topological decisions will be discussed.

1. Mokrani T., Scurrel M. Gas conversion to liquid fuels and chemicals: the methanol route-catalysis and processes development. Catalysis Reviews 2009, 51:1-145

2. de Klerk, A. Fischer-Tropsch Refining; Wiley-VCH Verlag & Co. KgaA: Weinheim 2011

3. Niziolek, A. M.; Onel, O.; Elia, J. A.; Baliban R. C.; Floudas, C. A. Coproduction of Liquid Transportation Fuels and C6_C8 Aromatics from Biomass and Natural Gas. AIChE Journal 2015, 61, 831-856.

4. Baliban, R. C.; Elia, J. A.; Floudas, C. A. Novel natural gas to liquids (GTL) technologies: Process synthesis and global optimization strategies. AIChE Journal 2013, 59, 505–531.

5. Baliban, R. C.; Elia, J. A.; Misener, R.; Floudas, C. A. Global Optimization of a MINLP Process Synthesis Model for Thermochemical Based Conversion of Hybrid Coal, Biomass, and Natural Gas to Liquid Fuels. Computers and Chemical Engineering 2012, 42, 64-86.

6. Niziolek, A. M.; Onel, O.; Floudas, C. A. Production of Benzene, Toluene, and Xylenes from Natural Gas via Methanol: Process Synthesis and Global Optimization. submitted

443544 Process Development for Ethanol and Isobutanol to Jet Fuel

Abstract:

Process Development for Ethanol and Isobutanol to Jet Fuels

Maoqi Feng, Cary Wong, Jimell Erwin and Darius Daruwalla, Chemical Engineering Dept., Division of Chemistry and Chemical Engineering, Southwest Research Institute, San Antonio, TX 78238.

Jet fuels production from renewable ethanol and isobutanol will significantly reduce the greenhouse gas emissions from airplanes and help the aviation industry less addict to fossil fuels. In this study, ethanol and isobutanol were converted to jet fuel boiling range hydrocarbons in a fixed-bed catalytic reactor. Figure 1 shows the block flow diagram. Three catalysts were evaluated, ZSM-5, MgO/alumina, and a large pore size zeolite, at three reaction temperatures: 300, 350, and 400°C. The per pass conversion yields were between 40-80%. Isobutanol gave higher conversion than ethanol. ZSM-5 had higher selectivity for aromatics such as mesitylenes, xylenes, and toluene. Catalyst lifetime study is still in progress.

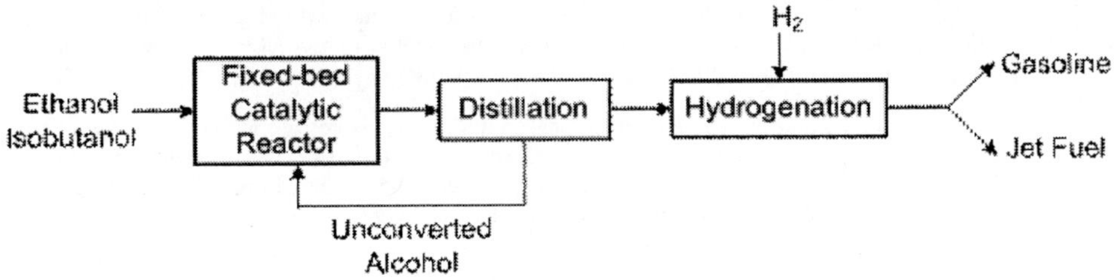

Figure 1. The Block Flow Diagram for the Alcohols to Jet Fuels Process

444081 Method for Hot Real-Time Sampling of Biomass-Derived Pyrolysis Vapors

Abstract:

The Thermochemical Process Development Unit (TCPDU) at the National Renewable Energy Laboratory (NREL) is a highly instrumented half-ton/day pilot scale plant capable of demonstrating industrially relevant thermochemical technologies from lignocellulosic biomass, including pyrolysis. Biomass derived pyrolysis vapors are a very complex mixture of over 400 known chemical components that undergo additional condensation reactions changing their composition as they are cooled and collected to form multiphasic oils. Added to this chemical change, up to 60% of the resulting products do not readily re volatilize, making off-line post analysis of condensed chemistry difficult. Real-time hot online sampling techniques, such as Molecular Beam Mass Spectrometry (MBMS), can provide real-time analysis providing operational indicators for performance.

It is difficult to obtain a representative real-time analytical sample of biomass derived pyrolysis vapors. Sampling must occur within a narrow range of temperature to reduce plugging of lines due to cooling and product composition changes due to overheating. Residence time within the sampling lines must be kept to a minimum to reduce further reaction chemistries. Solids from ash and char contribute to plugging and must be filtered at temperature. Pyrolysis vapors also form aerosols that are carried far downstream and can pass through filters resulting in build-up in downstream locations.

Experience at NREL has shown several key factors to consider when designing and installing an analytical sampling system for biomass pyrolysis vapors. They include minimizing sampling distance, effective filtering as close to source as possible, proper line sizing, even heating of all components, minimizing pressure drops, additional filtering or traps after pressure drops, and avoiding 90 degree bends to minimize vapor condensation and polymerization. Sample dilution may help with plugging but can actually cause aerosols to carry further downstream and will dilute trace analytes that may be of interest.

443719 Hydrotreating of Catalytically Upgraded Biomass Pyrolysis Oils

Abstract:

One way to make a liquid hydrocarbon fuel is fast pyrolysis of solid biomass followed by hydrotreating. Catalytic fast pyrolysis followed by hydrotreating is a potentially improved process which may reduce overall process costs by reducing the number of hydrotreating stages required, the amount of hydrogen consumed, and coking of the hydrotreating catalyst.

To assess the potential advantages of this method, three catalytic fast pyrolysis oils—produced with HZSM-5 at biomass-to-catalyst ratios from 0.5 to 2.1 and having oxygen contents from 3 to 18%—and a non-catalytic oil containing 28% oxygen, were hydrotreated in a batch autoclave with CoMo(S) at 330-420°C and 2500 psig total pressure.

Multiple catalytic-oil and temperature conditions produced products with oxygen contents about 1% by weight. Hydrogen consumption was less than ½ that of the non-catalytic oil. GC-MS analysis showed high concentrations of aromatics and hydrogenated aromatics. Phenols were the main oxygenated compounds detected in the products. The low-oxygen products and lower hydrogen consumption show that this is a promising alternative.

443542 Hydrodeoxygenation of Phenols and Derivatives to Aromatics

Abstract:

Phenols and substituted phenols can be produced from depolymerization of lignin, as we presented before at the AIChE conferences [1-2]. Hydrogenation of the phenols and substituted phenols to hydrocarbon aromatics provides a pathway for production of high octane number gasoline. In this study, phenol, anisole, 2,6-tert-butyl-3-methyl phenol and their mixtures were used as model compounds for hydrodeoxygenation catalyzed by different catalysts. The catalysts include conventional Ni(Co)Mo hydrotreating catalysts and noble metal catalyst supported on alumina. When toluene was used as the solvent, 100% conversion of the model compounds was obtained, however, methyl group transfer from the toluene to the model compounds were observed. The mechanism of the reactions is presented.

References

[1] Maoqi Feng, Darius Daruwalla, Kraft Lignin Depolymerization Under Hydrothermal Conditions, 2015 AIChE Spring Meeting & 10th Global Congress on Process Safety, Austin, Texas, April 27, 2015, Paper No. 56b.

[2] Maoqi Feng, Chee-Kai Tan and Darius Daruwalla, Lignin Depolymerization in Ionic Liquids with the Presence of Hydrogen Transfer Agents, 2014 AIChE Spring Meeting & 10th Global Congress on Process Safety, New Orleans, March 31-April 3, 2014, Paper No. 63d.

440851 Supercharge Styrene Polymerization Control with Dorf Ketal ACtify^{TM} Technology

Abstract:

Just over twenty years ago, the styrene monomer industry witnessed a dramatic change in the accepted method of in-process polymer control. In the 20 years since then, not a lot has changed. New chemistry, HPHA (HydroxyPropyl Hydroxylamine), was introduced with the intent to fully replace highly toxic, di-nitro retarders. However, limitations in effectiveness, particularly in the high temperature zones of the process made this impossible. Shortly thereafter, various forms of TEMPO compounds were offered, also with the expectation of being a "one product solution" for styrene polymerization control. HPHA and TEMPO were both highly successful at reducing the amount of steady-state polymer produced in a styrene distillation train. Unfortunately, because of a combination of economic and reliability concerns, most users decided to continue using Di-Nitro retarders in combination with the new inhibitors.

In the following years, many suppliers have tried to introduce new chemistries that would be safer to use and would serve the same functionality as a polymerization retarder. They have all failed to be accepted in the marketplace primarily for two reasons; They are too expensive, and they don't work as well as existing retarders.

A few years ago, researchers at Dorf Ketal discovered an amazing new chemical "Activator" that, when added to any existing inhibitor or retarder, dramatically improves the performance of the initial product. ACtify^{TM} inhibitors are able to maintain lower steady state polymer levels, even when used in lower doses. ACtify^{TM} retarders are able to maintain safe polymer levels for longer periods of time even when used at lower dosages. Users have achieved significant savings through reductions in yield loss and product usage. More importantly, this new ACtify^{TM} technology also works with non-toxic retarders. Instead of using hazardous di-nitro compounds, producers now have a safer option that is cost-effective, and actually works better! This paper will provide key information about the development and implementation of this new technology including data from commercial applications.

444016 Inherently Safer Design

Abstract:

Inherently Safer Design is a way of thinking differently from traditional hazard management. Instead of applying additional lawyers of safety to a process, the hazards may be eliminated or reduced through changes in materials, chemistry or process.

This paper will address the strategies of inherently safer design, the application of these strategies in projects, new technology facility modifications and changes in existing operations, how hazard and risk studies support these strategies, planning and conducting reviews.

Considerations for production targets, scheduling design and construction, location and separation of hazards, staffing, operating and maintenance and novel or never - tried technology ideas will be brain stormed and discussed while concluding this presentation.

444022 Adsorption Process for Removal of Phenol from Synthetic Waste Water By Using Activated Carbon

Abstract:

Mother Nature - we all have been talking about but how many of us are serious about protecting her. Already we have created havoc and one of the major contributors is chemical industries. Phenol is a refractory pollutant which is found in effluent. The concentration is low in the effluent but it is much effective to disturb the environment. Phenol can be removed economically by using cost effective techniques like adsorption.

Pilot plant lab scale set up was constructed. Groundnut shell was used as raw material to remove phenol from synthetic waste water. Activated carbon and charcoal were produced by pyrolysis process to carbonize the material and which were activated by using chemical activation process. Charcoal was treated with different acids such as HNO_3, H_2SO_4, HCL, H_3PO_4 and water for activation purpose and it was concluded that H_3PO_4 was the most suitable activating agent for groundnut shell charcoal.

The objective of this pilot plant experiment was to investigate the performance of charcoal and activated carbon as adsorbents for adsorbing phenol from its aqueous solution and to conclude their effectiveness and satisfactory results were obtained.

443281 Manage Risk and Improve Production Start-up of Capital Projects

Abstract:

In the final phases of a capital project, there are periods of high risk and several opportunities to shorten the time to full, on-spec production. Getting the control systems properly configured and optimized for stability is a critical path task. This presentation focuses on the effective use of control system information to rapidly achieve stable operations. It will also cover how to establish baselines for normal operation going forward.

Nonlinear partial least square (NPLS) methods with generalized likelihood ratio test (GLRT) for fault detection and diagnosis of chemical processes.

Chiranjivi Botre[a], Majdi Mansouri[b], Mohamed N. Nounou[c], Hazem N. Nounou[b] and M. Nazmul Karim[a]

[a] Artie McFerrin Dept. of Chemical Engineering, Texas A&M University, College Station, Texas 77843, USA
[b] Electrical and Computer Engineering Program, Texas A&M University at Qatar, Doha, QATAR,
[c] Chemical Engineering Department, Texas A&M University at Qatar, Doha, QATAR

1

Dwight Look College of ENGINEERING TEXAS A&M UNIVERSITY

16 AIChE Spring Meeting, Houston, TX

12TH GLOBAL CONGRESS ON PROCESS SAFETY

Why Fault detection (FD) ?

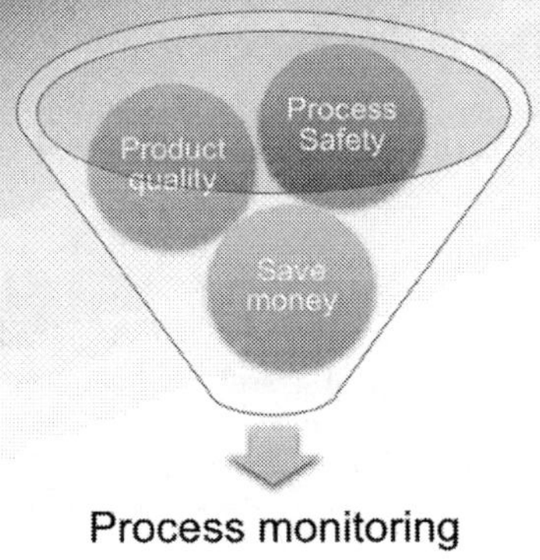

- Process faults are abnormal changes in several variables caused by deviation in process
- Need to develop monitoring techniques to detect accurately and quickly this abnormality in processes
- Fault detection is one of the important step in process monitoring

2

Software based fault detection methods

Model based
- First principle based model
- Accurate mathematical model required
- Difficult to derive model for complex chemical and environmental system

Knowledge (Qualitative) based
- Usually heuristic process
- Small scale systems with small variables
- Difficult to apply to complex systems

Data based
- Historical process data
- Can be applied to complex processes
- **Multivariate statistical methods (PLS)**

3

Dwight Look College of ENGINEERING TEXAS A&M UNIVERSITY
16 AIChE Spring Meeting, Houston, TX
12TH GLOBAL CONGRESS ON PROCESS SAFETY

Partial Least Square (PLS)

- Multivariate statistical methods are powerful tools capable of handling noise and highly correlated data sets.

regres

Process online variables such as temperature, flow rate, pressure etc.

Process quality variables such as concentration, product yield etc.

PLS model

X ⟷ Y

$$X = t_1 p_1^t + E_1 \qquad Y = u_1 q_1^t + F_1$$

$$Y = BX + \varepsilon$$

Where,

- t_1 and u_1 are the score matrix of X and Y respectively
- p_1^t and q_1^t are the loading vectors of X and Y respectively
- B is the regression coefficient

4

Iterative PLS algorithm (NIPALS)

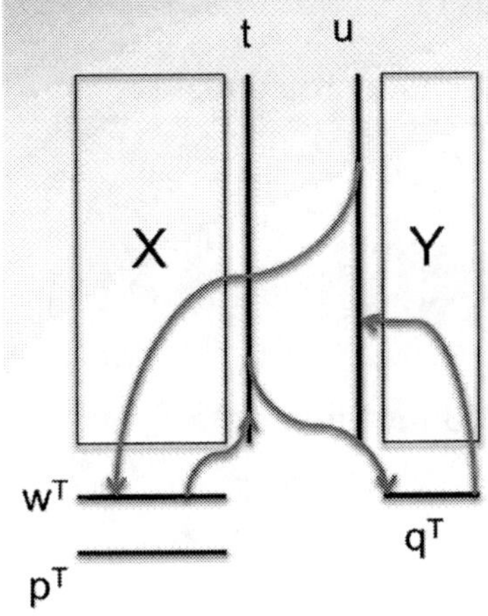

- The matrix X and Y are decomposed into latent structures using iterative methods.
- X is represented by its score t and Y is represented by u.
- For next iteration $$X_{k+1} = X_k - t_k p_k^T$$ $$Y_{k+1} = Y_k - t_k q_k^T$$
- Total latent variable $$\mathbf{T} = \begin{bmatrix} \mathbf{t}_1 & \mathbf{t}_2 & \cdots \\ \downarrow & \downarrow & \\ \vdots & \vdots & \end{bmatrix}$$

5

Dwight Look College of ENGINEERING TEXAS A&M UNIVERSITY 16 AIChE Spring Meeting, Houston, TX 12TH GLOBAL CONGRESS ON PROCESS SAFETY

FD Statistical tests

- **T square test**

$$T^2 = x^T \widehat{P}\widehat{\Lambda}^{-1}\widehat{P}^T x$$

Where:-
1. P is the loading vector matrix
2. Λ is the diagonal matrix of eigenvalues.

- **Q square test (SPE)**
  - ✓ Mean square error between X_{test_2} and transformed X_{test} from model

$$Q = \left\|\bar{x}\right\| = \left\|\left(1 - PP^T\right)X_{test}\right\|^2$$

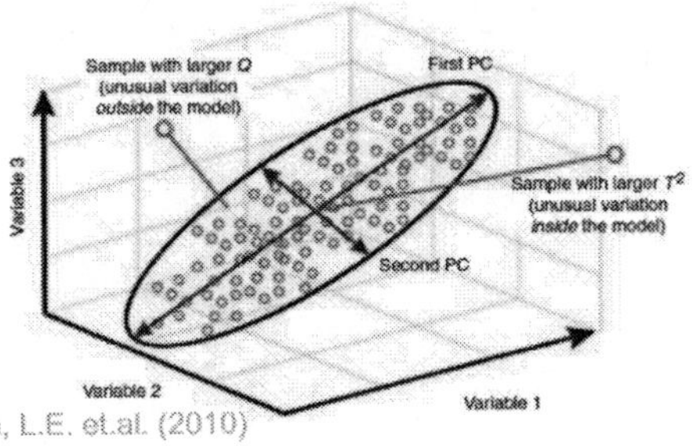

Mujica, L.E. et.al. (2010)

- T^2 statistics detects the variation with the subspace of the model which is higher then expected
- Q statistics detects the variation which is not explained by the model

Improved FD Statistical Test

Generalized likelihood ratio test (GLRT)

- The GLR test solves composite hypotheses testing problems by maximizing the likelihood ratio function over all faults.

$$\begin{cases} H_0 = \{E \sim N(0, \sigma^2 I_N)\} \\ H_1 = \{E \sim N(\theta, \sigma^2 I_N)\} \end{cases}$$

mean vector of fault and it is replaced by maximum likelihood estimate

- L(y) is the ratio of probability distribution of faulty data to faultless data

$$\delta(Y) = \begin{cases} H_0 & \text{if } L(y) \prec h(\alpha) \\ H_1 & \text{else.} \end{cases}$$

- Threshold value of GLRT is given by inverse of chi-square distribution

7

Overview of Process Monitoring

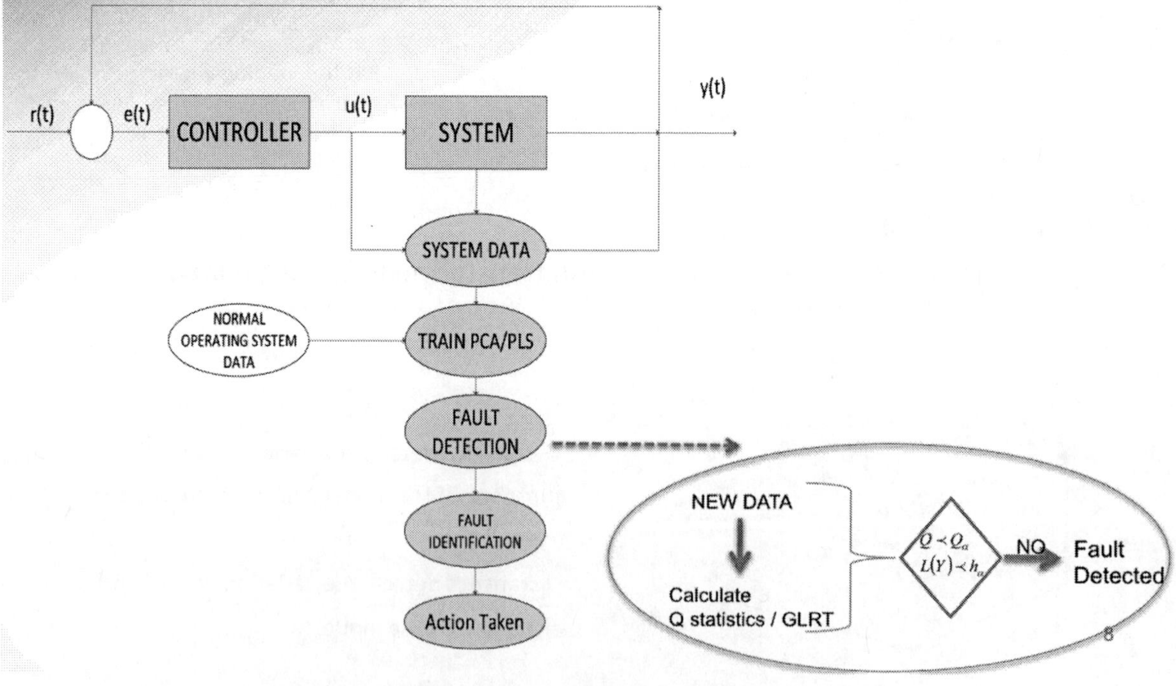

8

Dwight Look College of ENGINEERING TEXAS A&M UNIVERSITY
16 AIChE Spring Meeting, Houston, TX
12TH GLOBAL CONGRESS ON PROCESS SAFETY

Case Study: Tennessee Eastman Process (TEP) Problem

- Realistic simulation program of a chemical plant and is widely accepted as a benchmark for control and monitoring studies. It is a continuous process.
- The reaction scheme is given as

$$A(g) + C(g) + D(g) \longrightarrow G(liq)$$

$$A(g) + C(g) + E(g) \longrightarrow H(liq)$$

$$A(g) + E(g) \longrightarrow F(liq)$$

$$3D(g) \longrightarrow 2F(liq)$$

9

Dwight Look College of ENGINEERING TEXAS A&M UNIVERSITY
16 AIChE Spring Meeting, Houston, TX
12TH GLOBAL CONGRESS ON PROCESS SAFETY

TEP Process Flow Diagram

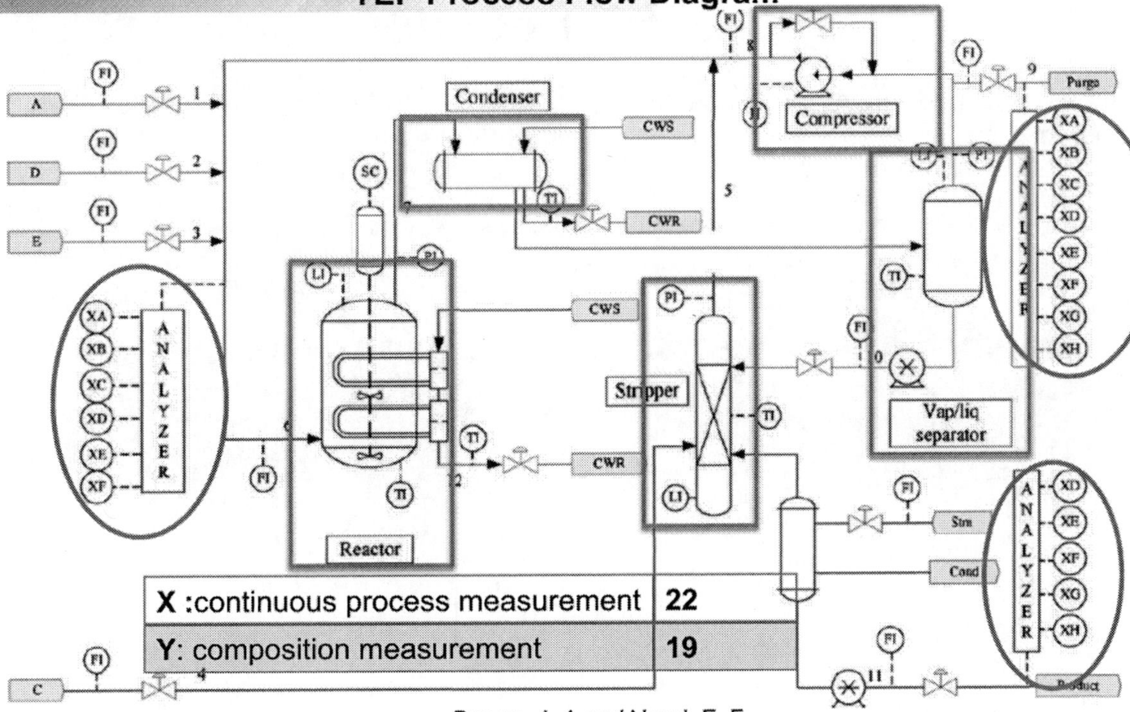

X :continuous process measurement	22
Y: composition measurement	19

Downs, J. J. and Vogel, E. F

Fault in Process Variables

Fault number	Process variable	Type
IDV(1)	A/C feed ratio, B composition constant	Step
IDV(2)	B composition, A/C ration constant	Step
IDV(3)	D feed temperature	Step
IDV(4)	Reactor cooling water inlet temperature	Step
IDV(5)	Condenser cooling water inlet temperature	Step
IDV(6)	A feed loss	Step
IDV(7)	C header pressure loss-reduced availability	Step
IDV(8)	A, B, and C feed composition	Random variation
IDV(9)	D feed temperature	Random variation
IDV(10)	C feed temperature	Random variation
IDV(11)	Reactor cooling water inlet temperature	Random variation
IDV(12)	Condenser cooling water inlet temperature	Random variation
IDV(13)	Reaction kinetics	Slow drift
IDV(14)	Reactor cooling water valve	Sticking
IDV(15)	Condenser cooling water valve	Sticking
IDV(16)	Unknown	Unknown
IDV(17)	Unknown	Unknown
IDV(18)	Unknown	Unknown
IDV(19)	Unknown	Unknown
IDV(20)	Unknown	Unknown
IDV(21)	The valve fixed at steady state position	Constant position

Yin et al. (2012)

- Total process time T=48 hrs.
- Fault was introduced at t=8 hrs.

11

GLRT Results

IDV 1: A/C feed ration, B composition disturbance

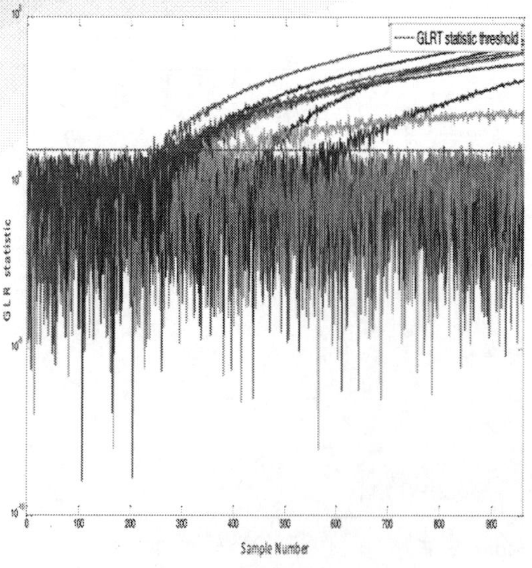

IDV 5: Condenser cooling water inlet temperature

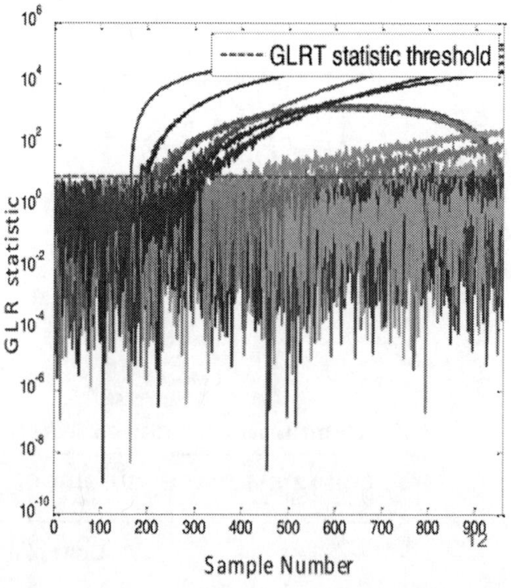

12

FD of Nonlinear Processes

- Most of the chemical processes are **nonlinear**, application of PLS-based monitoring to inherently nonlinear processes may lead to inefficient and unreliable process monitoring.

- Nonlinear PLS can be divided into two categories
  1. Observed X matrix of independent variables is projected onto a nonlinear surface
  2. Replacing the linear relation between the score vectors with a nonlinear form

13

Kernel PLS

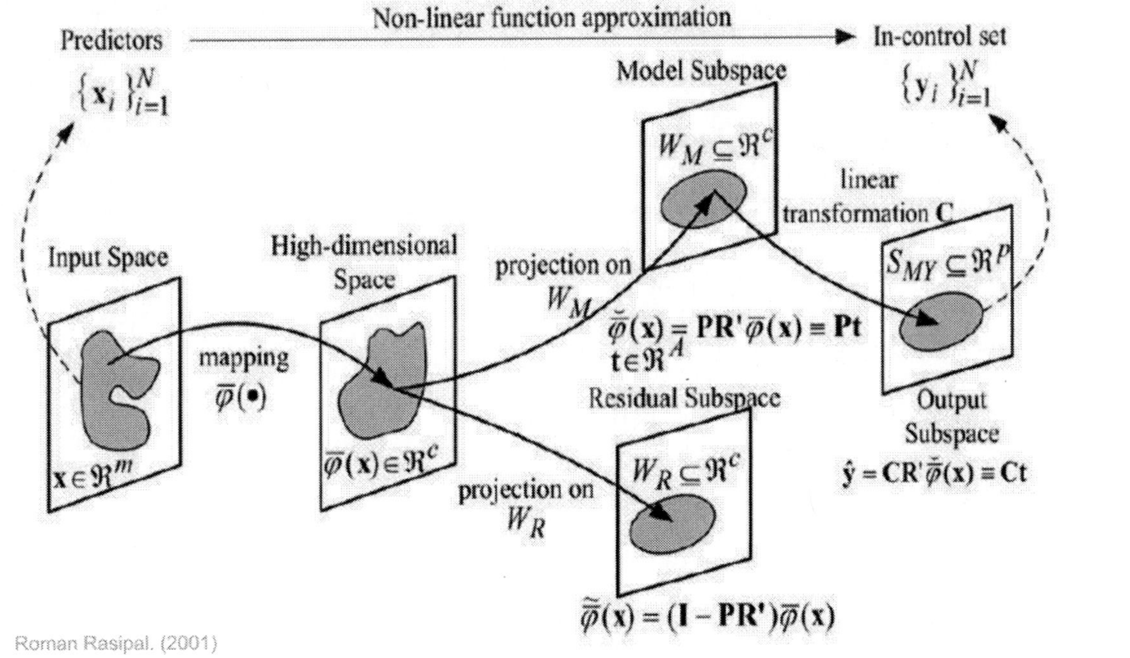

Roman Rasipal. (2001)

Kernel PLS

- Let Φ be a mapping from the original space Rd of the data vectors, into a inner product space F,

$$\Phi: \quad \chi \rightarrow F$$

- Mercer's Theorem shows that there exist a mapping into a space where a kernel function acts as a dot product.

$$k\left(x_i, x_j\right) = \Phi($$

- Kernel matrix K is defined as a symmetri matrix,

$$K_{ij} = k(x_i,$$

- The Gaussian radial basis function (RBF):

$$K(x,y) = \exp\left(-\frac{\|x-y\|^2}{c}\right)$$

- Polynomial Kernel:

$$K(x,y) = \langle x,y \rangle^d$$

- Sigmoid Kernel:

$$K(x,y) = \tanh\left(\beta_0 \langle x,y \rangle + \beta_1\right)$$

15

Dwight Look College of ENGINEERING TEXAS A&M UNIVERSITY
16 AIChE Spring Meeting, Houston, TX
12TH GLOBAL CONGRESS ON PROCESS SAFETY

Neural Network Partial Least Square (NNPLS)

- Linear relationship of input score matrix and output score matrix is replaced by neural net
- Neural net approximates the non linear function in this case
- Outer model is linear PLS and inside loop has neural net

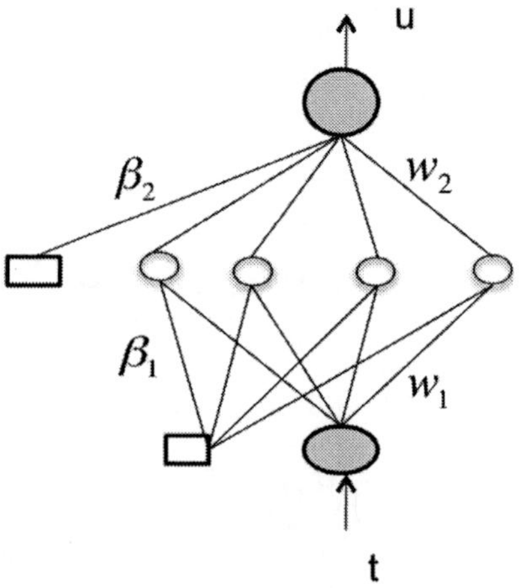

16

Neural network PLS

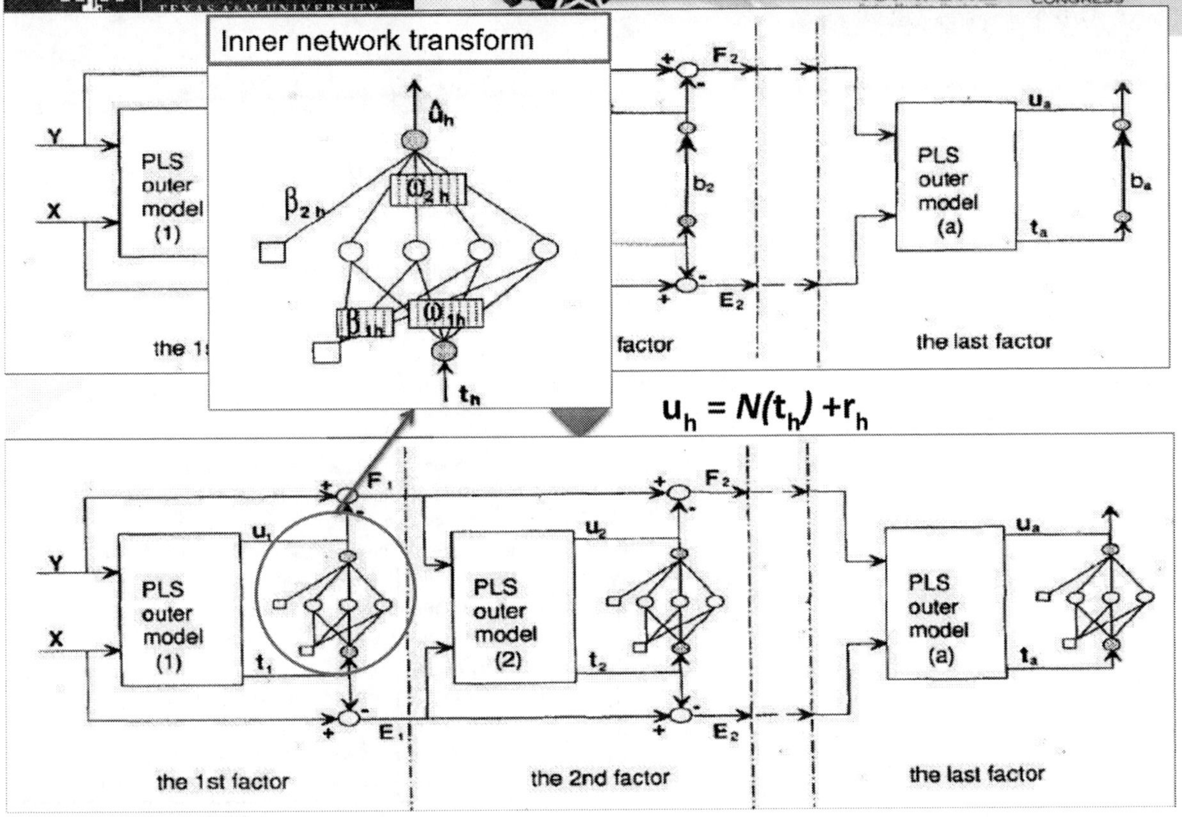

Dwight Look College of ENGINEERING TEXAS A&M UNIVERSITY

16AIChE Spring Meeting, Houston, TX

12TH GLOBAL CONGRESS ON PROCESS SAFETY

GLRT Results

IDV 1: A/C feed ration, B composition disturbance

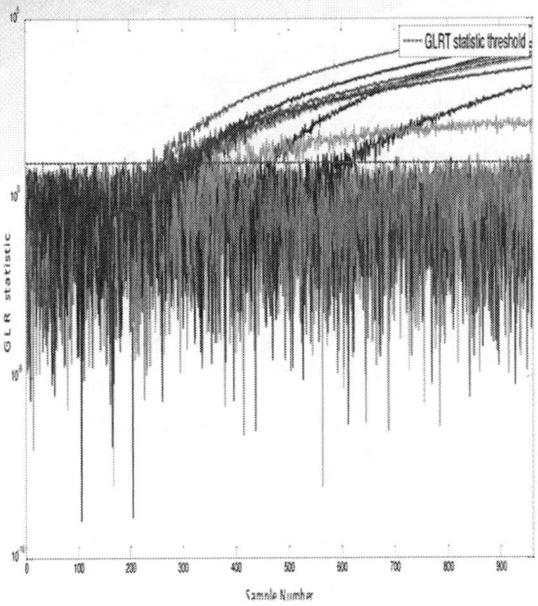

IDV 5: Condenser cooling water inlet temperature

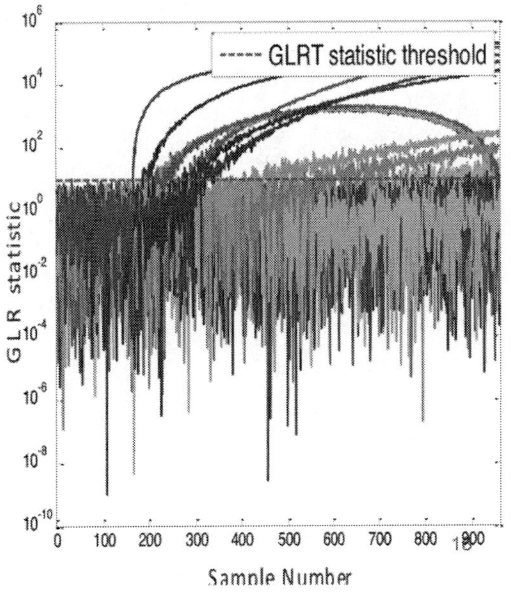

18

Dwight Look College of ENGINEERING TEXAS A&M UNIVERSITY

16 AIChE Spring Meeting, Houston, TX — 12TH GLOBAL CONGRESS ON PROCESS SAFETY

Contribution Plot

T = 166 sample

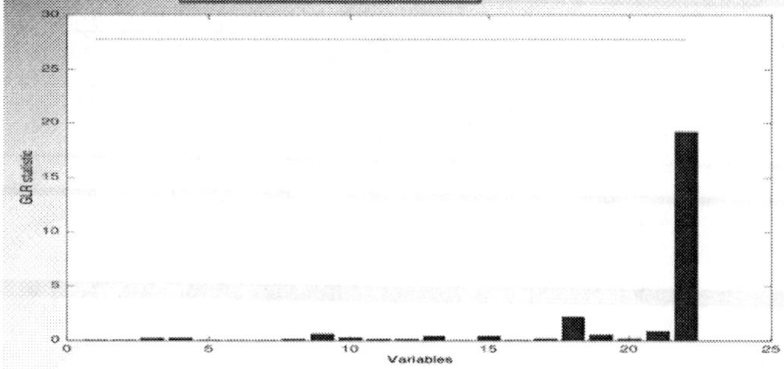

T = 168 sample

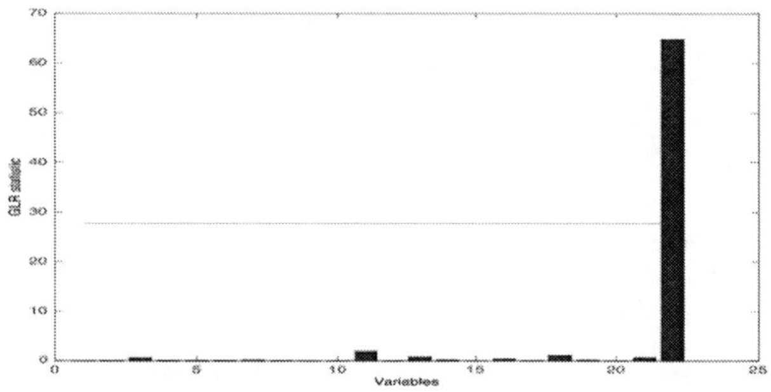

Process Variable	Description
XMEAS(1)	A Feed (stream 1)
XMEAS(2)	D Feed (stream 2)
XMEAS(3)	E Feed (stream 3)
XMEAS(4)	Total Feed (stream 4)
XMEAS(5)	Recycle Flow (stream 8)
XMEAS(6)	Reactor Feed Rate (stream 6)
XMEAS(7)	Reactor Pressure
XMEAS(8)	Reactor Level
XMEAS(9)	Reactor Temperature
XMEAS(10)	Purge Rate (stream 9)
XMEAS(11)	Product Separator Temperature
XMEAS(12)	Product Separator Level
XMEAS(13)	Product Separator Pressure
XMEAS(14)	Product Separator Underflow (stream 10)
XMEAS(15)	Stripper Level
XMEAS(16)	Stripper Pressure
XMEAS(17)	Stripper Underflow (stream 11)
XMEAS(18)	Stripper Temperature
XMEAS(19)	Stripper Steam Flow
XMEAS(20)	Compressor Work
XMEAS(21)	Reactor Cooling Water Outlet Temperature
XMEAS(22)	Separator Cooling Water Outlet Temperature

19

Dwight Look College of ENGINEERING TEXAS A&M UNIVERSITY

16 AIChE Spring Meeting, Houston, TX — 12TH GLOBAL CONGRESS ON PROCESS SAFETY

Contribution Plot

T = 300 sample

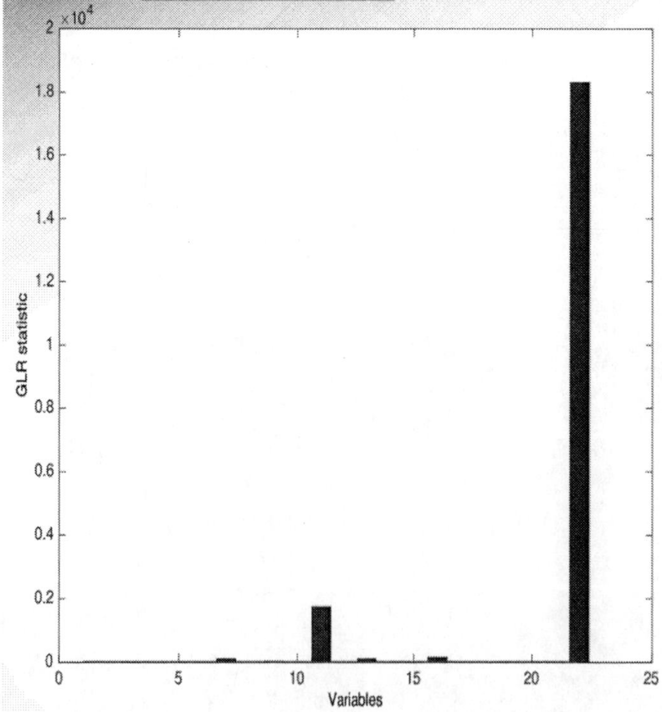

Process Variable	Description
XMEAS(1)	A Feed (stream 1)
XMEAS(2)	D Feed (stream 2)
XMEAS(3)	E Feed (stream 3)
XMEAS(4)	Total Feed (stream 4)
XMEAS(5)	Recycle Flow (stream 8)
XMEAS(6)	Reactor Feed Rate (stream 6)
XMEAS(7)	Reactor Pressure
XMEAS(8)	Reactor Level
XMEAS(9)	Reactor Temperature
XMEAS(10)	Purge Rate (stream 9)
XMEAS(11)	Product Separator Temperature
XMEAS(12)	Product Separator Level
XMEAS(13)	Product Separator Pressure
XMEAS(14)	Product Separator Underflow (stream 10)
XMEAS(15)	Stripper Level
XMEAS(16)	Stripper Pressure
XMEAS(17)	Stripper Underflow (stream 11)
XMEAS(18)	Stripper Temperature
XMEAS(19)	Stripper Steam Flow
XMEAS(20)	Compressor Work
XMEAS(21)	Reactor Cooling Water Outlet Temperature
XMEAS(22)	Separator Cooling Water Outlet Temperature

20

Dwight Look College of ENGINEERING TEXAS A&M UNIVERSITY

16 AIChE Spring Meeting, Houston, TX — 12TH GLOBAL CONGRESS ON PROCESS SAFETY

T = 400 sample

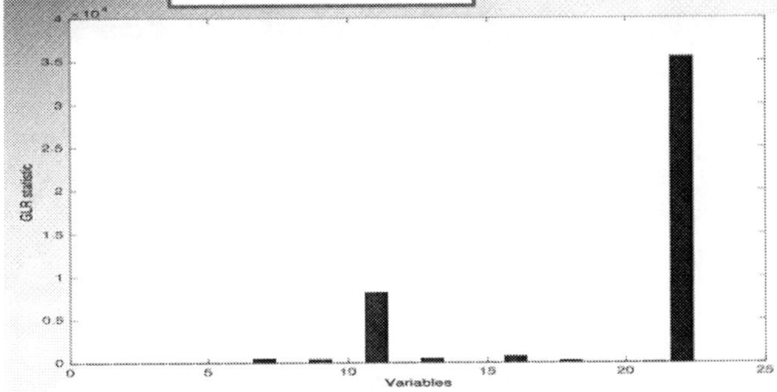

T = 850 sample

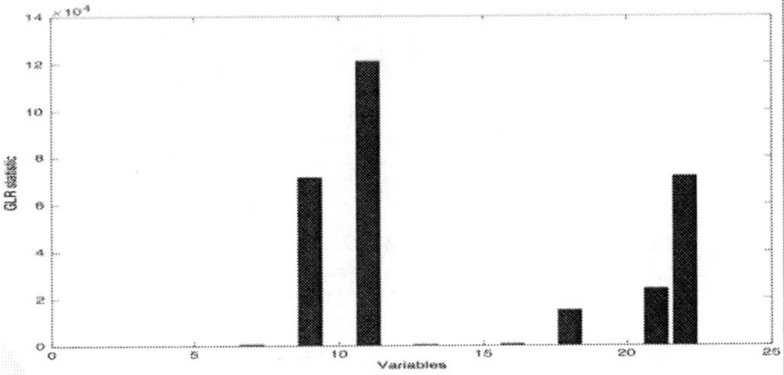

Contribution Plot

Process Variable	Description
XMEAS(1)	A Feed (stream 1)
XMEAS(2)	D Feed (stream 2)
XMEAS(3)	E Feed (stream 3)
XMEAS(4)	Total Feed (stream 4)
XMEAS(5)	Recycle Flow (stream 8)
XMEAS(6)	Reactor Feed Rate (stream 6)
XMEAS(7)	Reactor Pressure
XMEAS(8)	Reactor Level
XMEAS(9)	Reactor Temperature
XMEAS(10)	Purge Rate (stream 9)
XMEAS(11)	Product Separator Temperature
XMEAS(12)	Product Separator Level
XMEAS(13)	Product Separator Pressure
XMEAS(14)	Product Separator Underflow (stream 10)
XMEAS(15)	Stripper Level
XMEAS(16)	Stripper Pressure
XMEAS(17)	Stripper Underflow (stream 11)
XMEAS(18)	Stripper Temperature
XMEAS(19)	Stripper Steam Flow
XMEAS(20)	Compressor Work
XMEAS(21)	Reactor Cooling Water Outlet Temperature
XMEAS(22)	Separator Cooling Water Outlet Temperature

21

Dwight Look College of ENGINEERING TEXAS A&M UNIVERSITY

16 AIChE Spring Meeting, Houston, TX — 12TH GLOBAL CONGRESS ON PROCESS SAFETY

Regression Model

- Predicted Y is given as

$$Y = \beta X_{new} + \varepsilon$$

- The regression coefficient is given by

$$\beta = W\left(P^T W\right)^{-1} Q^T$$

Weight W and loading vector P and Q

$$W = X^T U$$

$$P = X^T T\left(T^T T\right)^{-1}$$

$$Q = Y^T T\left(T^T T\right)^{-1}$$

- Regression Coefficient

$$\beta = X^T U\left(T^T X X^T U\right)^{-1} T^T Y_{train}$$

22

Dwight Look College of ENGINEERING TEXAS A&M UNIVERSITY

16 AIChE Spring Meeting, Houston, TX

12TH GLOBAL CONGRESS ON PROCESS SAFETY

KPLS Regression Model

- Kernel regression coefficient is defined as

$$\beta = \varphi_{new}^{T} U\left(T^{T} K^{T} U\right)^{-1} T^{T} Y_{train}$$

Kernel matrix of training data

$$K = k\left(x_{train}, x_{train}\right) = \varphi\left(x_{train}\right)\varphi\left(x_{train}\right)$$

Kernel matrix of new data set

$$K_t = k\left(x_{new}, x_{train}\right) = \varphi\left(x_{new}\right)\varphi\left(x_{train}\right)$$

- Output predicted from KPLS

$$\widehat{Y} = K_t U\left(T^{T} K^{T} U\right)^{-1} T^{T} Y_{train}$$

23

Dwight Look College of ENGINEERING TEXAS A&M UNIVERSITY

16 AIChE Spring Meeting, Houston, TX

12TH GLOBAL CONGRESS ON PROCESS SAFETY

TEP Process Flow Diagram

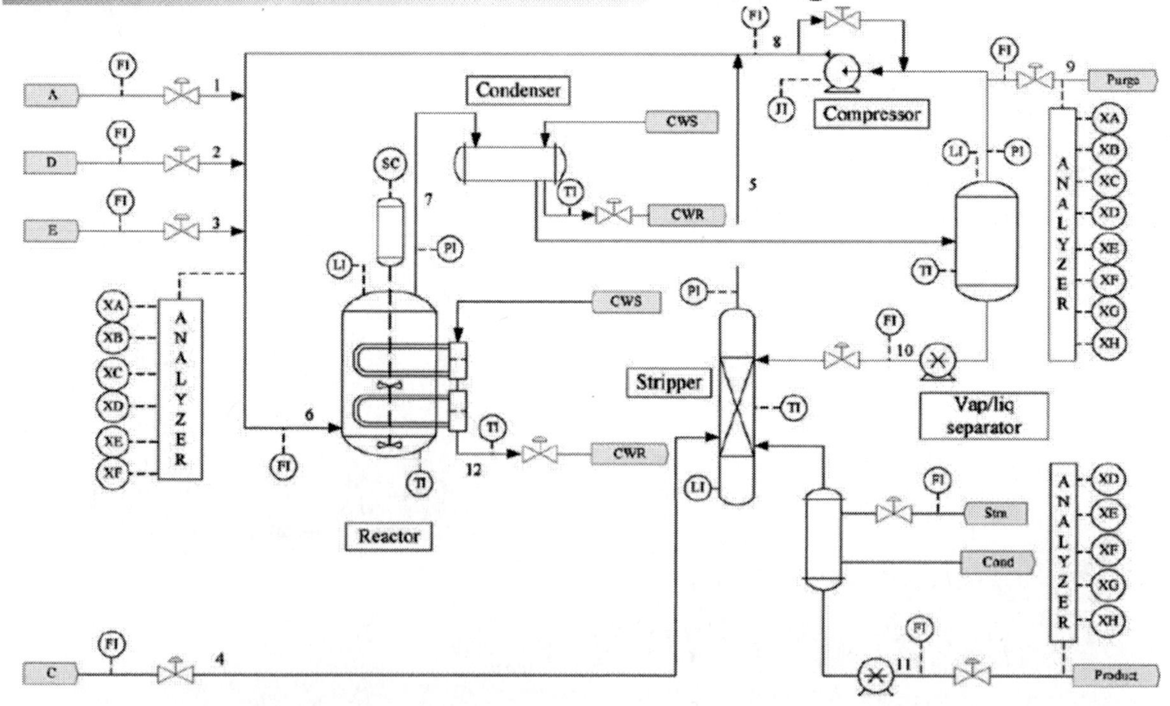

CONTINOUSLY MONITORED PROCESS VARIABLES

Process Variable	Description	Units
XMEAS(1)	A Feed (stream 1)	kscmh
XMEAS(2)	D Feed (stream 2)	kg/hr
XMEAS(3)	E Feed (stream 3)	kg/hr
XMEAS(4)	Total Feed (stream 4)	kscmh
XMEAS(5)	Recycle Flow (stream 8)	kscmh
XMEAS(6)	Reactor Feed Rate (stream 6)	kscmh
XMEAS(7)	Reactor Pressure	kPa Gauge
XMEAS(8)	Reactor Level	%
XMEAS(9)	Reactor Temperature	°C
XMEAS(10)	Purge Rate (stream 9)	kscmh
XMEAS(11)	Product Separator Temperature	°C
XMEAS(12)	Product Separator Level	%
XMEAS(13)	Product Separator Pressure	kPa Gauge
XMEAS(14)	Product Separator Underflow (stream 10)	m^3/hr
XMEAS(15)	Stripper Level	%
XMEAS(16)	Stripper Pressure	kPa Gauge
XMEAS(17)	Stripper Underflow (stream 11)	m^3/hr
XMEAS(18)	Stripper Temperature	°C
XMEAS(19)	Stripper Steam Flow	kg/hr
XMEAS(20)	Compressor Work	kW
XMEAS(21)	Reactor Cooling Water Outlet Temperature	°C
XMEAS(22)	Separator Cooling Water Outlet Temperature	°C

25

Dwight Look College of ENGINEERING TEXAS A&M UNIVERSITY
16 AIChE Spring Meeting, Houston, TX
12TH GLOBAL CONGRESS ON PROCESS SAFETY

KPLS Regression Plots

$$A(g) + C(g) + D(g) \longrightarrow G(liq)$$

$$A(g) + C(g) + E(g) \longrightarrow H(liq)$$

$$A(g) + E(g) \longrightarrow F(liq)$$

$$3D(g) \longrightarrow 2F(liq)$$

Regression plot of product F

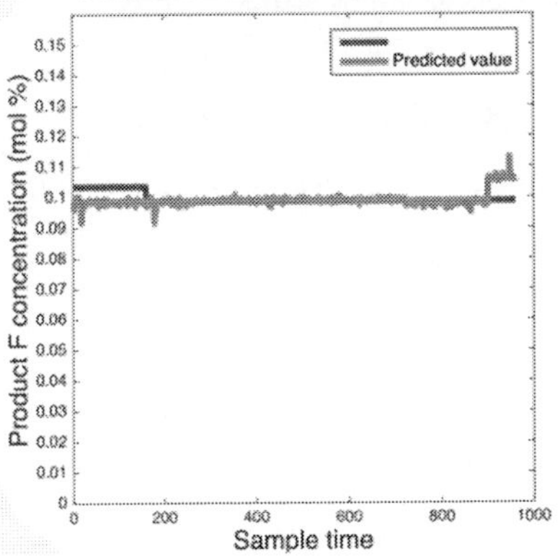

Regression plot of product G

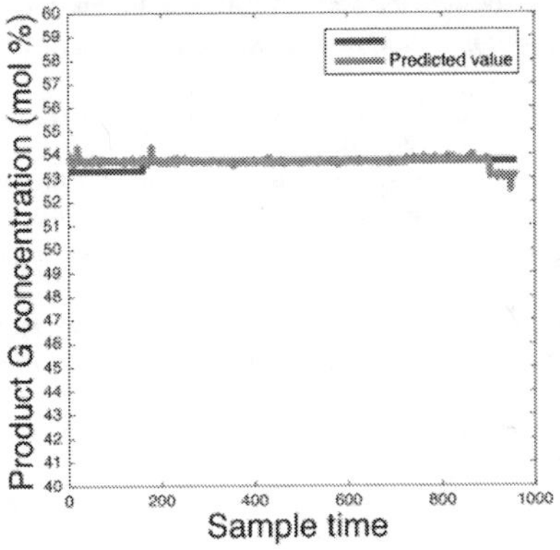

Regression plot of product H

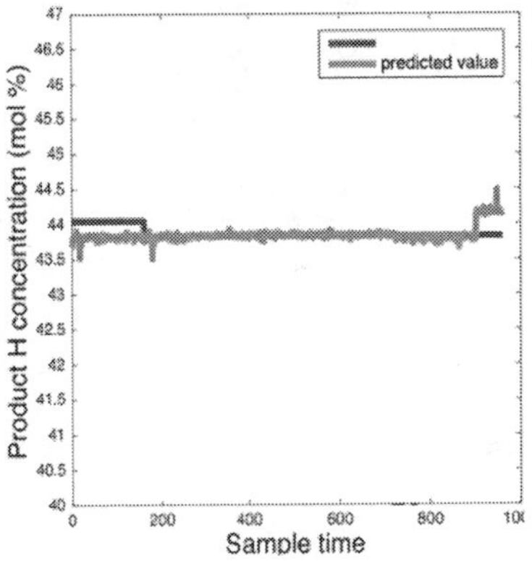

Summary

- We presented the problem of fault detection using kernel partial least square (PLS) -based generalized likelihood ratio test (GLRT) and neural net partial least square (PLS) –based GLRT.
- TEP results demonstrate the effectiveness of the KPLS -based GLRT technique for detection of multiple faults with low false alarm rate and early fault detection
- KPLS regression model is used to predict concentration of the product from online process variable.

27

Dwight Look College of ENGINEERING TEXAS A&M UNIVERSITY
16 AIChE Spring Meeting, Houston, TX
12th GLOBAL CONGRESS on PROCESS SAFETY

References

- Harrou Fouzi, Mohamed N. Nounou, Hazem N. Nounou and MudduMadakyaru, "Statistical fault detection using PCA -based GLR hypothesistesting" Journal of Loss Prevention in the Process Industries 26 (2013) 129e139
- Hodge, D., L. Simon, and M.N. Karim. "Data Driven Approaches to Modeling and Analysis of Bioprocesses: Some Industrial Examples." In American Control Conference, 2003. Proceedings of the 2003, 3:2062–2076 vol.3, 2003.
- Kourti Theodora and John F. Macgregor. "Process Analysis, Monitoring and Diagnosis, Using Multivariate Projection Methods." Chemometrics and Intelligent Laboratory Systems 28 (1995) 3-21
- Kourti, Theodora, Paul Nomikos, and John F. MacGregor. "Analysis, Monitoring and Fault Diagnosis of Batch Processes Using Multiblock and Multiway PLS." *Journal of Process Control*, IFAC Symposium: Advanced Control of Chemical Processes, 5, no. 4 (August 1995): 277–84.
- MacGregor, J. F., and T. Kourti. "Statistical Process Control of Multivariate Processes." *Control Engineering Practice* 3, no. 3 (March 1995): 403–14.
- Mansouri, M. M., Nounou, H. N., Nounou, M. N., Datta, A. A., 2014. State and parameter estimation for nonlinear biological phenomena modeled by s-systems. Digital Signal Processing 28, 1–17, 2014.
- Qin S. Joe and T.J. McAvoy. "Nonlinear PLS modeling using neural networks." Computers them. Engng, Vol. 16, No. 4, pp. 379-391, 1992
- Reyes, L.H., Gomez, J.M., and Kao, K.C. Improving carotenoids production in yeast via adaptive laboratory evolution. Metabololic Engineering. 21 (2014) 26-33.
- Rosipal, Roman "Nonlinear Partial Least Squares: An Overview." *Chemoinformatics and Advanced Machine Learning Perspectives: Complex Computational Methods and Collaborative Techniques*, 2010, 169–89.
- S. Joe Qin, "Statistical process monitoring: basics and beyond" Journal of Chemometrics 2003; 17: 480–502.
- Yoon, Seongkyu, and John F. MacGregor. "Principal-component Analysis of Multiscale Data for Process Monitoring and Fault Diagnosis." AIChE Journal 50, no. 11 (November 1, 2004): 2891–2903. doi:10.1002/aic.10260.
- Zhang, Yingwei, and S. Joe Qin. "Improved Nonlinear Fault Detection Technique and Statistical Analysis." AIChE Journal 54, no. 12 (December 1, 2008)
- Zhang, Yingwei, Wenyou Du, Yunpeng Fan, and Lingjun Zhang. "Process Fault Detection Using Directional Kernel Partial Least Squares." *Industrial & Engineering Chemistry Research* 54, no. 9 (March 11, 2015): 2509–18.

28

Thank You

Questions?

443424 Energy Consumption Reduced By 8% in an Aromatic Extraction Plant

Abstract:

The costs related with energy in a basic supplies unit are the second largest expenditure surpassed only by raw material. Facing the intense international competitiveness, any effort to reduce the energy consumption helps to improve the business performance. The enhancement of an energy efficiency indicator, besides the economical benefits to the company, improves the energetic performance, which is a comparison standard for petrochemical industries.

Taking this scenario into account, an Advanced Process Control (APC) system was implemented in the benzene, toluene and xylene extraction plant in UNIB2-RS/Braskem unit. The project's main objective was to achieve and sustain a lower energy consumption, covering all extraction plant. As result, the APC obtained a reduction in the solvent flow, used in the components separation, by 10% generating an 8% cut in steam consumption. This reduction will guarantee an economy estimated in R$ 900 thousand (U$200,000) per year.

State Based Control and the Value Delivered from the Initial Design Through the Operating Life of the Facility

Tom Nolan
ProSys
2822 O'Neal Lane Bldg A, Baton Rouge LA 70816 USA
Tom.Nolan@prosys.com
with Stephen Bihl, Chris Woolfley, Craig Leopard,
Stacey Kevin, Will LaPointe, Brian White, Stephen Villarrubia,
Daniel Forman, Jared Bullock

Prepared for Presentation at
American Institute of Chemical Engineers
2016 Spring Meeting
11th Global Congress on Process Safety
Houston, Texas
April 10-14, 2016

UNPUBLISHED

State Based Control and the Value Delivered from the Initial Design through the Operating Life of the Facility

Tom Nolan
ProSys
2822 O'Neal Lane Bldg A, Baton Rouge LA 70816 USA
Tom.Nolan@prosys.com

Keywords: State Based Control, Procedural Automation, Knowledge Transfer, Instrument Justification, Alarm Rationalization

Abstract

This paper examines the attributes of automation through state based control and the value delivered to manufacturing from the initial design through the operating life of the facility. It examines both the benefits of improving a non-state based control facility to state based control, as well as the front end loading of green field project with state based control incorporating instrument and alarm justification into the design process for improved performance and reduced cost. Through knowledge capture and transfer into the control system, this paper demonstrates how state based control will give you, essentially, the best operator on the board performance all of the time. The paper also examines how state based control enhances a traditional single loop safety system with a coordinated safety response. The coordinated safety response manages all outputs of a given unit to a safe state. Communication to other units allows them to remain at their highest state of readiness to return to normal operation reducing process interruption.

1. Introduction

It should be noted that this paper discusses the value that automation with state based control can add to a facility from initial design through the operating life of the facility. This is in no way meant to imply that state based control is the only way to achieve these objectives. As will be seen, the work process for state based control gives an organized way of adding value in the design and commissioning phases. State based control also provides a solid frame work in which to hang the automation and gives a convenient way to communicate between units. That is to say that the same level of automation can probably be achieved with or without state based control, but the frame work provided by "states" makes is much easier to accomplish, maintain and improve. As will be seen, it also improves the operability of the control.

There are two types of state based control:

- Inferred State – Using process measurements to infer what state the process is in.
- Driven State – Using the DCS to drive the outputs to put the process in the desired state.

2. Knowledge Capture and Transfer

Throughout the life cycle of the facility, knowledge capture and transfer are crucial to successful and safe operation. This process should begin in the early front-end loading (FEL) stages. Sometimes, plants are not started up, commissioned, or operated exactly the way they were designed to be. The involvement of control engineers, process engineers and equipment subject matter experts can capture, during the initial FEL, how the equipment is designed to be operated. This has a tremendous impact on safety and loss prevention, especially when bringing in new equipment or even a new technology that operations personnel are not familiar with. The automation should help operations in a challenging situation rather than confuse the issues at hand with alarm showers. Unproven equipment, operators new to running it and alarm showers are a very bad combination. When equipment is started up according to automated procedures, with no shortcuts, the dynamic alarm management is going to work better and last longer with fewer unplanned events (see the section on the effect of unplanned events). This is a great time to capture the knowledge of the design team and transfer it into the automation where it is always available for use and continuous improvement throughout the life time of the facility.

As experience is gained in the plant through the years of operation, learning is going to occur. Sometimes this is learned the hard way - through incidents or near misses. Of course, the knowledge can be captured in procedures, but they can also be captured in automation and automated procedures. This builds the knowledge transfer into the DCS system. Through the operating life of the facility, the automation can become smarter and more robust.

Loss of process knowledge through attrition is often an issue for those facilities that do not have a high level of automation or state based control. A large percentage of the work force is nearing retirement age. Within the next five years, about 20% of the workforce could retire. In a 2012 report, the Social Security Administration estimated 10,000 will retire per day. The younger workforce is going to be smaller and more mobile. Younger workers consider strategic job-hopping normal. According to the Bureau of Labor Statistics, the average worker stays at each job for 4.4 years. Most millennials expect that they will change jobs every three years or less. For large continuous plants that may not be long enough to even see one turnaround. John Grubbs, author of "Surviving the Talent Exodus", describes them as the nomadic generation. He points out that only companies that can accelerate the time it takes to make a young worker competent will benefit. They need to become competent quickly enough to be productive until they move on to another opportunity. Knowledge capture and transfer into the automation of the facility can have a huge impact. This knowledge transfer has to happen before the experienced personnel retire.

All of the above knowledge capture can be leveraged to new green field projects with similar process technology through Standard Reusable Architecture. This increases the return on investment in the automation.

3. Level of Automation

The level of automation needs to be determined early on in the front-end loading. There are a number of considerations in determining what to automate and what level of automation to shoot for.

Part of this will come from the alarm justification process. There are ISA 18.2 metrics regarding operator response. You should keep this in mind when considering the span of control of the operator and the required level of automation. In the determination of Causes, Consequences and Actions, if the required response time from an operator is not realistic, consider automating the response or accepting the consequences that have been identified.

Veronica Lindström and Mats Winroth list some considerations (ref 2):

1. Increase labor productivity
2. Reduce labor cost
3. Mitigate the effects of labor shortages
4. Reduce or eliminate routine manual or clerical tasks
5. Improve worker safety
6. Improve product quality
7. Reduce manufacturing lead time
8. Accomplish processes that cannot be done manually
9. Avoid the high cost of not automating

Consider that operator errors result in the highest average dollar loss per major incident at over $80 million (J. H. Marsh & McLennan, 2010). Also, according to the ACR Advisory Group, operator error accounts for 42% of the unscheduled plant shutdowns with equipment failures and process upsets rounding out the list (O'Brien, 2010).

Historically, trust has been an issue with the level of automation (ref 3) and always will be. In the past, it was more "what do we trust automation to do", and that is moving toward "what do we trust not being automated" more and more. Certainly, we would want to automate safety shutdown logic with people less and less willing to take credits for operator actions. It is probably not "a matter of time" before we have self-driving cars as much as we will get to the point where we look back and think "it was crazy and dangerous back when we use to drive our own cars around".

It cannot be overstated - simulation for check out and training is crucial to properly working software working properly and operators confidence in automation. With that trust, operators move from manipulating each small part of the process to managing the process with their heads up and getting the big picture. (ref 1,3)

4. State Based Control Adds Value in Early Front End Loading

Instrumentation is expensive. It has to be purchased, installed, maintained, added to the graphics and configured in the DCS. An instrument is likely to end up with more alarms on it than are needed when its configured in the DCS. This could be because it is easier to add alarms to the instrument than it is to figure out if they are required or not, or just because somebody along the way thinks it should be alarmed. This is the point where alarm rationalization needs to take place. The alarms that are needed should be a factor in determining the instrumentation list. Here are the instruments, now we need to put alarms on them is not the way to go.

5. Instrumentation Needs to be Justified

If the instrumentation is not deliberately justified for use early in the front-end loading, it causes problems. You may end up with instruments that are not needed, causing unnecessary expense and cost of ownership. Worse yet, instruments that should be there and are not, lead to operability issues. Adding these instruments later on in the project adds much more expense and rework into the project. Before the end of FEL-2, you should have a good list of the instruments that are needed to take forward to the cost estimate. If justified for use, there should be minimal creep through the process.

Given the total cost to install and the cost of ownership, start from the Process Flow Sheets with no instrument. Then add instruments only as the use for the instrument is justified for the appropriate level of automation and alarming. Consider instruments needed for automated start up / shutdown, control, alarms, safety instrumented system, environmental, troubleshooting, redundancy for reliability, and accounting. Then move on to the P&IDs and minimize reworking them.

Below is a simple example showing instrumentation justified for use in each process state.

Figure 1: Justify the Instrumentation for Use through Each State

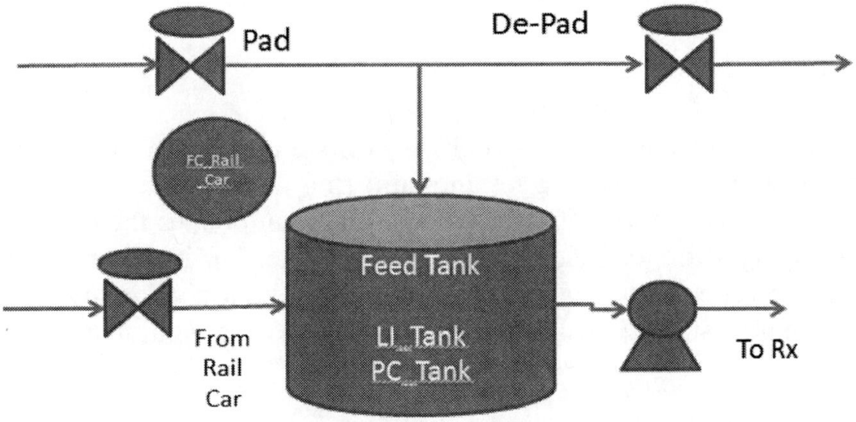

IO List	Justification
LI_Tank	Safety / Cntl
PC_Tank	Safety / Cntl
CV_DePad	Cntl
CV_Pad	Cntl
FC_Rail_Car	Cntl / Acct
CV_Rail_Car	Cntl

State	PVH_LI_Tank	PVH_PC_Tank	PC_Tank	FC_Rail_Car	PVL_LI_Tank
Maintenance	Emergency	High	DePad		
Manual Line Up	Emergency	High	Depad		
Fill	Emergency	High	Pad / DePad	On / Off Level	Low
Run	Emergency	High	Pad / DePad		High
Run and Fill	Emergency	High	Pad / DePad	On / Off Level	High

For example, working through bringing the process up in the maintenance state, a high level and high pressure alarm would be needed. Needing a level and pressure alarm justifies the need for the level and pressure instruments. Working through to the fill state, a control valve to fill and a flow transmitter would be required. The flow transmitter would also be used for accounting purposes. Working through starting the unit up in each case instruments are added as justified creating the instrument list and the control narrative. With this process, instruments are not added that are not needed, and instruments that are needed are not missed causing either poor performance or increased cost and rework.

6. State Based Control Simplifies the Math for Increased Levels of Automation

As we well know, control programs can become complex with higher levels of automation. Put simply, a well-designed state based control can simplify the process making it easier to design, create and operate at higher levels of automation.

Think of an algebraic equation Like $X^3 + 6 X^2 + 11 X + 6$. This equation can be factored into $(X + 3) * (X^2 + 3X + 2)$, and further factored into $(X + 3) * (X + 2) * (X + 1)$. This can give simpler equations to work with and understand. If the states of a process are designed well, they can be used as factors that simplify the Boolean logic and make it more robust for automation. So whatever logic was required to get the unit to the state it is in does not need to be repeated. All of the checks and balances to get the process to that state have been done, and any logic that is in place to move the process to another state are in place as well if the process situation changes. To simplify the code, the states Boolean factors out the logic that can be used. This will be seen as very useful in the coordinated safety response. Once you identify a safe state and how the outputs should be managed in that safe state, the only decision is "when do you take the unit to the safe state". Let any credited safety function be performed in the safety system and let the DCS respond accordingly in response to the safety system action.

7. State Based Enablement

An "enable" is fundamentally a decision. It is like a switch to turn something on or off. An enable can let a controller mode be able to operate, or let an alarm ring if the conditions come true. It might also indicate that flow is needed to a control module that manipulates the outputs. In state based control, alarms are pretty strongly based on the process state. For instance, a low level alarm would not be enabled in a process state where the level should be low, like down for maintenance, but enable in the running state. The high level alarm however would probably always be enabled.

Please note that in the discussion below for logic that $\sum$ will represent a Boolean summation with "Or" logic and π will represent a Boolean product with "And" logic applied through the brackets. So:

$$\sum(A, B, C) = A \text{ Or } B \text{ Or } C$$
$$\pi(A, B, C) = A \text{ And } B \text{ And } C$$

For example a distillation tower may have a number of states in which the steam valve should be open. A first pass at the steam valve logic with respect to states might be:

BV_Steam = State_Heat_Up Or State_Total_Reflux Or State_Recycle Or State_Run

That may not be sufficient in that other state dependent conditions might need to be applied in some but not all of the states.

For example if the operator should verify the manual line up of the steam system before the steam valve is allowed to operate in the heat up state.

BV_Steam = (State_Heat_Up And Line_Up_Confirmation) Or State_Total_Reflux Or State_Recycle Or State_Run

There may be an interlock that applies to all of the states. So the equipment is not operated unless the interlock is false.

For example don't open the steam valve if there is a high level or temperature no matter what state you are in.

Steam_Valve_Interlock = PVH_Tower_Level or PVH _Tower_Temperature

BV_Steam = [(State_Heat_Up And Line_Up_Confirmation)
Or State_Total_Reflux Or State_Recycle Or State_Run]
And Not Steam_Valve_Interlock

Note that the interlock is easily maintained by adding or removing conditions. For instance, adding that there is not a high delta P in the tower. The frame work is there.

There may also be instances where a permit is required, which would be a condition that must be true regardless of state to operate the valve.

For example the cooling system is seen to be in operation because a fan is on.

Steam_Valve_Permit = Fan_Running

BV_Steam = [(State_Heat_Up And Line_Up_Confirmation)
Or State_Total_Reflux Or State_Recycle Or State_Run]
And Steam_Valve_Permit
And Not Steam_Valve_Interlock

Why an interlock and a permit? Since the Interlock and Permit act on all of the states, this could be accomplished with just one variable that acts on all the state instead of two. The reason for two is it makes the logic easier to write, troubleshoot and maintain. For interlocks, it is easier to write the "true" logic for the equipment cannot operate. The logic is typically $\sum$(Condition_i). For permits, they are typically like a daisy chain of condition that must come true in order to indicate that it is ok to operate, π(Condition_i). Both of the lists are easily modified as need be.

All of the above combined gives a basic Unified State Based Control Enablement Theorem :

Permit = π(Condition_To_Allow_Operation_In_Any_State_i)

Interlock = ∑(Condition_To_Interlock_Operation_In_Any_State_i).
Enable = [∑(State_i) And π(State_Dependent_Conditions _i)]
And Permit
And Not Interlock

Where it is understood that some states may not have a special condition for that state and a Permit or Interlock may, or may not be needed.

The above equations can be used in state based control as an organized robust means of enabling many things based on the state.

The above form of the enablement equation can be used to open a steam valve as mentioned above or enable alarms or controller modes. It can also be used to trigger sequences in control or equipment modules. For instance – it can calcuate when a fuel gas flow is required and send the signal to a double block and bleed module. Upon receiving the signal for flow or stopping flow it will sequence the bleed, upstream and downstream valves. Likewise, a pump with a suction, motor and discharge valve module could receive an instruction that flow is required or not required and in the same way sequence the suction, motor, and discharge outputs to achieve flow or stop flow.

8. Unit Communication

Within a process, cell units provide services to other units and require services from other units to make the product. For instance, a raw material feed tank may provide feed to a reactor as a service, but requires the thermal oxidizer to treat vents from the tank's pad/de-pad system. This becomes very important in properly starting up the process as well as normal shutdowns, emergency shutdowns and degradations where the process is not shutdown, but not 100% where you would want it to be.

In a similar fashion to the state based enablement, Boolean logic can be constructed based on the states to signal when a unit needs, is requesting, or can provide a service to the process. This may be all that is needed or could go to something like the ISA 88 States.

For instance, a Boolean can be created in the Thermal Oxidizer unit communicating that it can provide vent treatment. Within each unit that requires the thermal oxidizer in the state transition logic, the Thermal_Oxidizer_Can_Provide_Vent_Treatment flag is required to move to a state where it will create vents that need to be treated. Each unit that requires vent treatment can construct a similar Boolean based on the states that it needs vent treatment. For instance, Feed_Tank_Needs_Vent_Treatment. Within the thermal oxidizer, the transition logic will check that no units need vent treatment before it transitions out of a state where it can provide it, unless it is required for a safety trip. The safety trip scenario will be discussed in the degradation scenario section.

There are several advantages to the Booleans based on state. One is that individual developers may not know that much about other units that they are not working on. For instance, if the guy doing the feed tank doesn't know that much about the thermal oxidizer, he knows that he needs it and that by convention the developer will create a flag when it can treat vents. Likewise, the developer of the thermal oxidizer knows that all users by convention will create a flag that they need vent treatment.

Another advantage to the communication Booleans is that they enforce a normal proper start up and shutdown of the process. The thermal oxidizer has to come up first before the units that require vent treatment can come up, and they must go down first before the thermal oxidizer can come down. This will eliminate being able to come up without environmental devices that are required by the permit to operate the facility. This prevents environmental issues, fines or loss of license to operate.

They provide an easy way to identify and deal with degradation scenarios.

9. Degradation Scenarios

If things work normally, the units will come up and down as needed based on when they can provide or need services. What happens when a service is needed but cannot be provided, as in the case of a thermal oxidizer trip? For instance, the feed tank would not be allowed to transition to a state where vent treatment was needed without it being available. After it is up and running, what happens when the vent treatment service can no longer be provided? If something is needed, but cannot be provided it is a degradation scenario.

The state based control can be structured to keep the process at its highest state of readiness, given the degradation, so that it can return to normal operation as quickly as possible.

So in the case of the above mentioned degradation, depending on the mechanical layout of the plant and what the plant is permitted to do, the solution that causes the least problems can be implemented when the degradation is detected. For instance, if the operating permit allows venting to the atmosphere until the thermal oxidizer can be restarted, the system could vent for a short time. However, if venting is not permitted and a flare header is available, then vent to the flare header.

Since each unit can communicate what it needs and what can be provided, a matrix of degradation scenarios can be created to always keep the plant at the highest state of readiness. As another example, if the reactor that is providing feed to the finishing section trips it can move to a total reflux or recycle state until the reactor comes back online.

As an example, consider a process with a feed tank, a reactor, a distillation tower, and a thermal oxidizer. In order to keep the example brief, just a few services and scenarios are shown (not intended to be comprehensive).

Figure 2

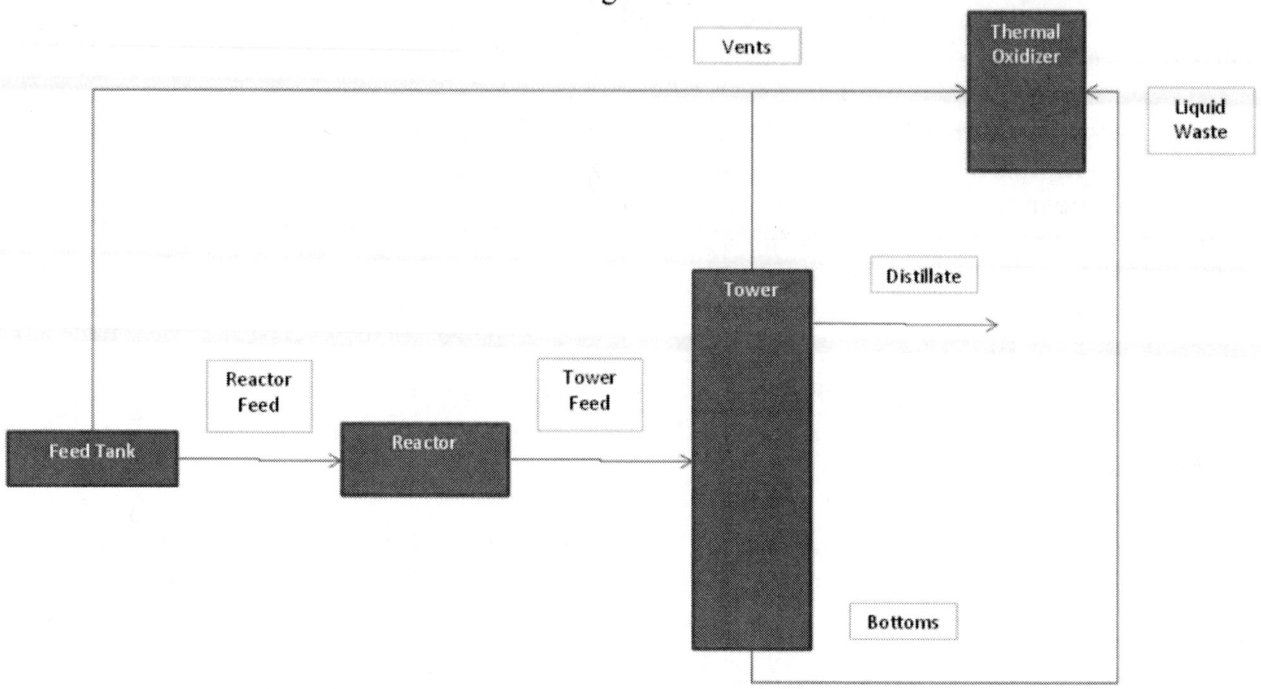

Table 1. Services Provided

Unit	Service Provided 1	Service Provided 2	Service Provided 3
Feed Tank	Provide Feed to the Reactor		
Reactor	Taking Feed From the Feed Tank	Provide Feed to the Tower	
Tower	Taking Feed From the Reactor	Providing Distillate	Providing Bottoms
Thermal Oxidizer	Vent Treatment	Liquid Waste Treatment	

Table 2. Services Used

Unit	Service Used 1	Service Used 2	Service Used 3
Feed Tank	Vent Treatment		
Reactor	Feed From Feed Tank		
Tower	Feed From the Reactor	Vent Treatment	Liquid Waste Treatment
Thermal Oxidizer			

Since it is known that degradation occurs if we lose something that is being provided, a degradation matrix can be created to give the required actions. Please note that the action do not necessarily have to be automated. It is a better idea to figure out what needs to happen, and then decide which things will be automated.

Table 3. Degradation Matrix

	Thermal Oxidizer Can't Treat Vents	Thermal Oxidizer Can't Treat Liquid Waste	Feed Tank Can Not Provide Reactor Feed	Reactor Can Not Feed the Tower	Tower Can't Take Feed From The Reactor
Feed Tank	Switch Vents to the Flare Header		XXXXX	Stop Feeding the Reactor	Stop Feeding the Reactor
Reactor			Flush With Solvent and Swap To Solvent Recovery	XXXXX	Flush With Solvent and Swap To Solvent Recovery
Tower	Switch Vents to the Flare Header	Switch Tower Bottoms to the Burn Tank (If burn tank level become too high go to total reflux)		Go to Total Reflux	XXXXX
Thermal Oxidizer	XXXXX	XXXXX			

This exercise is very useful in discovering operability issues early. For instance, with the "Thermal Oxidizer Can't Treat Liquid Waste" degradation, if there is not a Burn Tank in the process design, the tower would have to go to total reflux and stop production in the plant. The plant production could have been hanging on a flame detector. In general, a detailed study of using automation to bring the plant up can uncover start up issues and avoid the "we have to build it running, because there is no way to start it up" syndrome.

History trending the communication Booleans can be very beneficial for troubleshooting as well since they will indicate when degradation have occurred.

10. Coordinated Safety Response

A trip from the safety system will in all probability result in a degradation scenario. A piece of equipment that was needed to provide something is gone. With this degradation recognized in the control system, the process can automatically remain at the highest possible state of readiness under the current circumstances.

The safety system without state based control is designed well and tested to achieve the desired probability of failure on demand. The safety instrumented function will protect against the scenario for which it was designed. It will also cause many operability problems that have to be dealt with by the operator manually when it activates. What follows in trying to deal with the trip may inadvertently lead to the next trip / incident.

For instance, if a SIF cuts steam to a reactor to prevent an over pressure situation the reactor is probably not going to over-pressure and rupture. But what else has to happen to deal with the trip? You are probably in a high stress situation, maybe in high stress situation, such as an alarm shower. Do you stop the feeds to the reactor? Do you leave one feed running to prevent polymerization and cut another one off to stop the exotherm? Do you go to full cooling? Do you vent to the flare header? Would doing any of those things cause more problems? In the midst of this, the back end of the plant has just lost feed and it will all have to be dealt with manually while getting the reactor squared away.

It would be a lot better to have the most experienced operator on the board, when he is well rested and having a good day. A less experienced board operator may have never seen this scenario and there is not time to go get the procedures. Even the most experienced board operator may be fatigued or under a lot of stress.

With automation, you not only have the best board operator on all the time, but you also have the cumulative knowledge from the design team along with what has been learned about the plant through start up and commissioning and operation to this point.

11. Standard Reusable Architecture

To save cost and improve quality whenever possible, embrace the use of a standard reusable architecture. "State based control has been implemented and standardized for some time through ISA S88. It is often associated with batch processes but works equally well for continuous processes. S88 was written to improve efficiency. It provides a standard way to specify equipment and control modules in an architecture that can be re-used. ISA 106 Procedure Automation for Continuous Process Operations is in development."(ref 1). For more detailed information on the ISA 88 physical model see, ISA 88.

On a very high level, a unit module can be thought of as something on the order of a unit operation, like a reactor for instance. The architecture can be structured such that the unit module is made up of equipment modules. An equipment module might be on the order of a charging system. The charging system may be a grouping of a number of different things like valves, pumps, screw conveyers, etc. that work together to provide the reactor charge. The final actuators of the equipment module may be in control modules that actually interface with the IO. So, the equipment module may contain a number of control modules. The control module implements the basic control.

Once investments are made in the development of Control, Equipment and Unit Modules, they can be reused as applicable. For instance, a reactor to make a product may need to feed in a particular reactant to a given amount for the batch size, at a rate, with certain constraints. To make the product, these things are not going to change. There may be a facility in a large integrated site that has a header that can provide the reactant. In another site, it may be stored in a tank. In another site, it might be put on totes and weighed in.

With proper consideration to architecture, a unit module can be built for the reactor that is isolated from the variability in the instrumentation by control modules, essentially driving as much variability as possible into the control modules. So, the unit reactor module can be used for any application of the reactor. Control and equipment modules can be built to deliver the flow of reactant according to the parameters specified from the reactor unit module. There may be reuse of the control and equipment modules as well. For instance, the tote may have application in other places.

In terms of skill set, the reactor is going to require a great deal of process knowledge and process control expertise. Building control and equipment modules to deliver an amount of flow at a given rate will probably take a little less expertise. The instantiation of previously created unit, equipment and control modules will probably take even less expertise. So in effect, this leverages the process and process control knowledge.

12. Impact on Unplanned Events

Unplanned events in a manufacturing facility are expensive and dangerous. "Some analysis has been done linking unplanned events to alarm rates and the level of automation of a facility, with the facilities divided into four quadrants." (ref 1)

As you might expect quadrant one with low alarm loading and high automation was the least likely to have unplanned events. Quadrant 4 with high alarm loading and low automation was the most likely to be have unplanned event.

With the recent emphasis on alarms, it should be noted that comparing Quadrant 2 with high alarm loading and a high level of automation to Quadrant 3 with low alarm loading and a low level of automation that Quadrant 2 had less unplanned event than quadrant 3.

Figure 3: Relative Unplanned Events by Quadrant

Quadrant 2 High Alarms High Automation	**Quadrant 4** High Alarms Low Automation
Quadrant 1 Low Alarm High Automation	**Quadrant 3** Low Alarms Low Automation

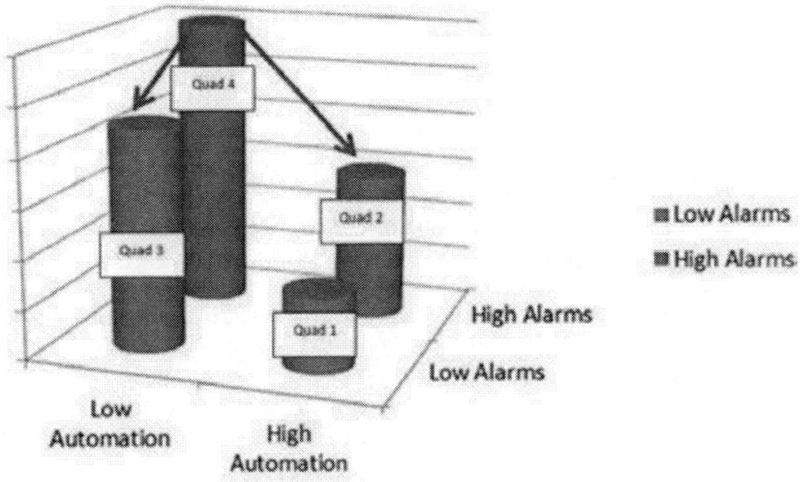

It can be seen from the slope on the surface of the alarm loading axis that alarms are important, but the surface on the automation axis has twice the slope of the alarm axis in the reduction of unplanned events

This is a very significant opportunity. Alarm management is important, but automation can have twice the impact.

13. Conclusion

Addressing automation through state based control from the early front-end loading stages can be a systematic way of justifying instruments and dynamic alarm management. It can also reduce cost and improve performance of a facility. State based control gives you the ability to capture and transfer knowledge (from the original design team through start up and commissioning and then ongoing run time) in order to keep equipment operating as it was designed, for safer more reliable and economical operation. The plant can get smarter with time and that knowledge can be embedded in the control system for whoever is on the board regardless of experience level. It is never too late to start. Automation can be added in as a consideration from the beginning of the facility or added in at any time. A reusable architecture will increase the return on investment in automation while improving quality and consistency.

Operators are in a position to manage the process through state changes while having their heads up to see the big picture - avoiding problems and optimizing performance. Managing alarms is important in the reduction of unplanned events. Automation can have an even larger impact on reducing unplanned events than alarm management. Bottom line alarm management and automation are both things that need to be addressed for a safe productive plant.

State based control allows project teams to meet and exceed the rising expectations they are challenged with.

References

[1] 1.0 Change Your Board Operator to a Process Manager with State Based Control (Tom Nolan with Katherine Persac)

[2] 2.0 Aligning manufacturing strategy and levels of automation: A Case Study (Veronica Lindström and Mats Winroth)

[3] http://www.diva-portal.org/smash/get/diva2:355948/fulltext01.pdf

[4] 3.0 Determining Appropriate Levels of Automation FITS SRM Automation Management Research (Charles L. Robertson)

[5] https://www.faa.gov/training_testing/training/fits/research/media/Det_App_Lvl_Atm.pdf

[6] http://www.controlglobal.com/articles/2003/181/

[7] https://www.isa.org/isa88/

Fault Detection Using Multiscale Shewhart Charts: Algorithm and Applications

M. Ziyan Sheriff
Texas A&M University, College Station, TX

M. Nazmul Karim Mohamed Nounou

Overview

- Introduction
  - Univariate Process Monitoring
  - Conventional Shewhart Chart
- Multiscale Representation of Data
- Multiscale Shewhart Chart
- Applications
- Conclusion & Future Directions

1

TEXAS A&M UNIVERSITY TEXAS A&M UNIVERSITY at QATAR

Process Monitoring

- What is process monitoring?
  - Observation of a particular process parameter to ensure that it stays within set control limits.
- Why process monitoring?
  - To ensure safe operation
  - To maintain product quality

2

TEXAS A&M UNIVERSITY TEXAS A&M UNIVERSITY at QATAR

Process Monitoring

- Based on data being used:
  - Quantitative methods
  - Qualitative methods
  - Process-history (data) based methods
- Based on the monitored variables:
  - Univariate methods
  - Multivariate methods

3

Univariate Data-based Monitoring Methods

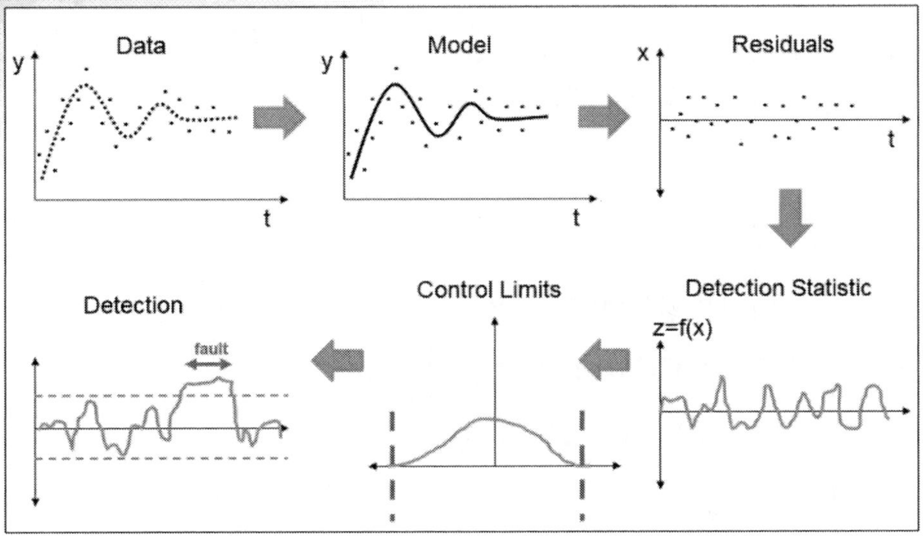

Fig. 1: Application of Univariate Charts for monitoring

4

Shewhart Chart

ENGINEERING
TEXAS A&M UNIVERSITY

TEXAS A&M UNIVERSITY | TEXAS A&M UNIVERSITY at QATAR

Shewhart Chart

- Developed by Walter Shewhart (1920s) working for Bell Systems [1]
- Main Features
  - Center (Targeted process mean)
  - Upper Control Limit (UCL)
  - Lower Control Limit (LCL)

[1] W. A. Shewhart, "Quality control charts," *Bell Syst. Tech. J.*, vol. 5, no. 4, pp. 593–603, 1926.

5

TEXAS A&M UNIVERSITY | TEXAS A&M UNIVERSITY at QATAR

Shewhart Chart

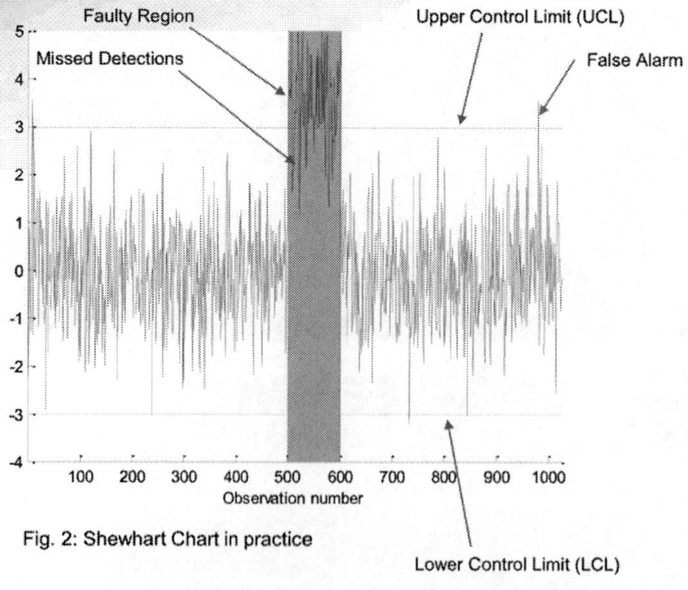

Fig. 2: Shewhart Chart in practice

Performance evaluated based on missed detection and false alarm rates

6

Assumptions of Shewhart Chart

1. Residuals contain a moderate level of noise
2. Residuals are independent (uncorrelated)
3. Fault-free residuals are normally distributed (Gaussian)

7

Wavelet-Based Multiscale Representation of Data

ENGINEERING
TEXAS A&M UNIVERSITY

Multiscale Representation: Schematic

Low-pass filter: $\phi_{ij}(t) = \sqrt{2^{-j}}\phi(2^{-j}t - k)$

High-pass filter: $\psi_{ij}(t) = \sqrt{2^{-j}}\psi(2^{-j}t - k)$

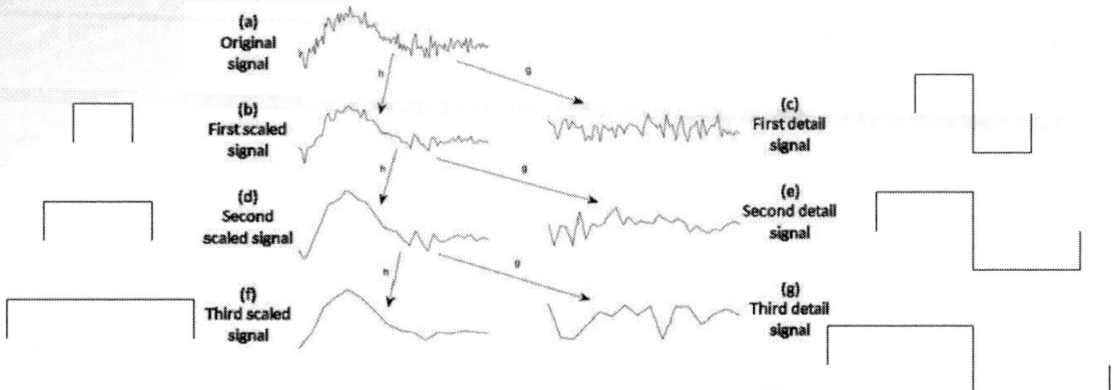

Fig. 3: Schematic to display Multiscale Representation [2]

Original signal: $x(t) = \sum_{k=1}^{n2^{-J}} a_{Jk}\phi_{Jk} + \sum_{j=1}^{J}\sum_{k=1}^{n2^{-j}} d_{jk}\psi_{jk}(t)$

[2] M. Z. Sheriff, F. Harrou, and M. Nounou, "Univariate process monitoring using multiscale Shewhart charts," in 2014 International Conference on Control, Decision and Information Technologies (CoDIT), 2014, pp. 435–440.

Advantages of Multiscale Representation

1. Ability to naturally denoise signals

2. Decorrelation of autocorrelated data at multiple scales

3. Wavelet coefficients (detail) are Gaussian regardless of distribution of time-domain data

TEXAS A&M UNIVERSITY | TEXAS A&M UNIVERSITY at QATAR

1. Naturally Denoise Signals

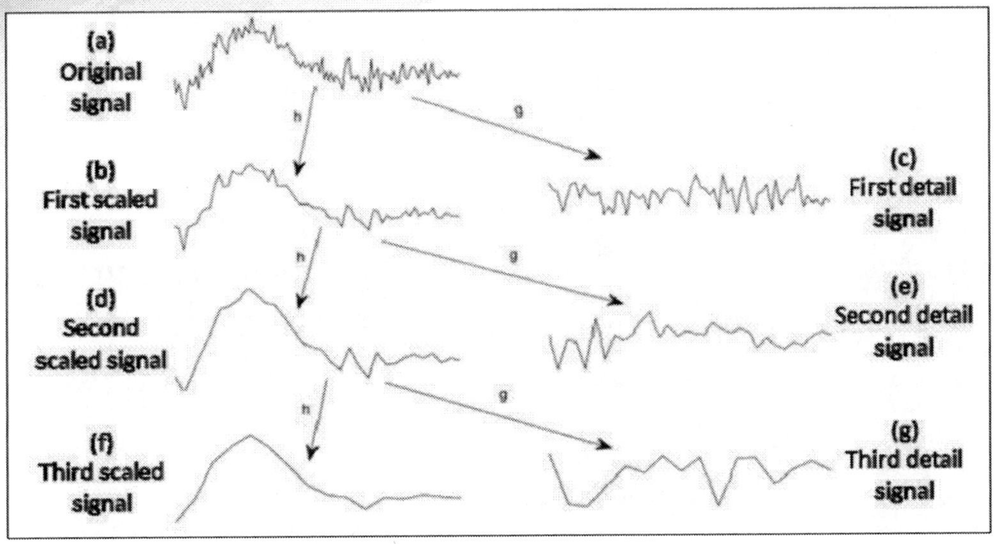

Fig. 4: Schematic to display Multiscale Representation

10

TEXAS A&M UNIVERSITY | TEXAS A&M UNIVERSITY at QATAR

2. Decorrelation of Autocorrelated Data

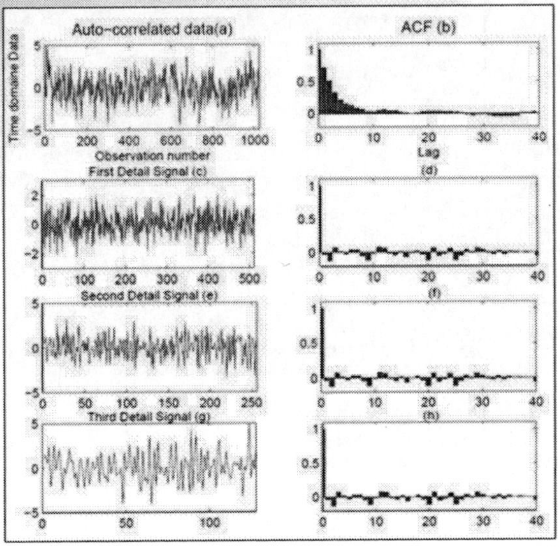

Decorrelation of autocorrelated data at multiple scales

Fig. 5: Decorrelation of autocorrelated AR(1) data at multiple scales.

11

Data are Gaussian at Multiple Scales

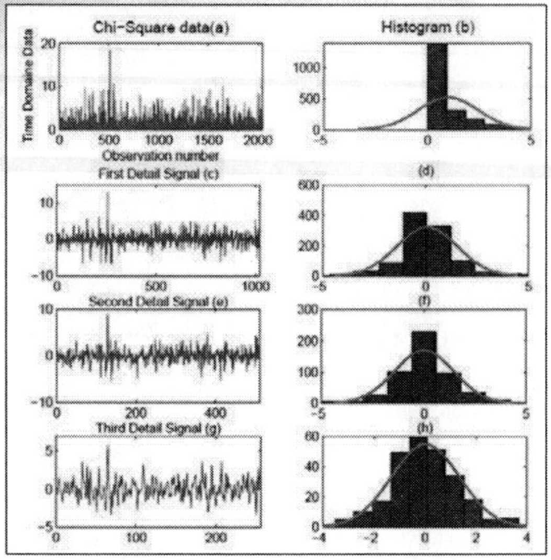

Fig. 6: Distribution of chi-square data at multiple scales.

Wavelet coefficients (detail) are Gaussian regardless of distribution of time-domain data

12

Multiscale Shewhart Chart Algorithm

ENGINEERING
TEXAS A&M UNIVERSITY

TEXAS A&M UNIVERSITY | TEXAS A&M UNIVERSITY at QATAR

Multiscale Algorithm: Main Idea

- Apply Shewhart chart on the detail signals
- Details signals better satisfy the underlying assumptions
- Create a simplified multiscale Shewhart chart for fault detection

13

TEXAS A&M UNIVERSITY | TEXAS A&M UNIVERSITY at QATAR

Multiscale Shewhart Chart Algorithm

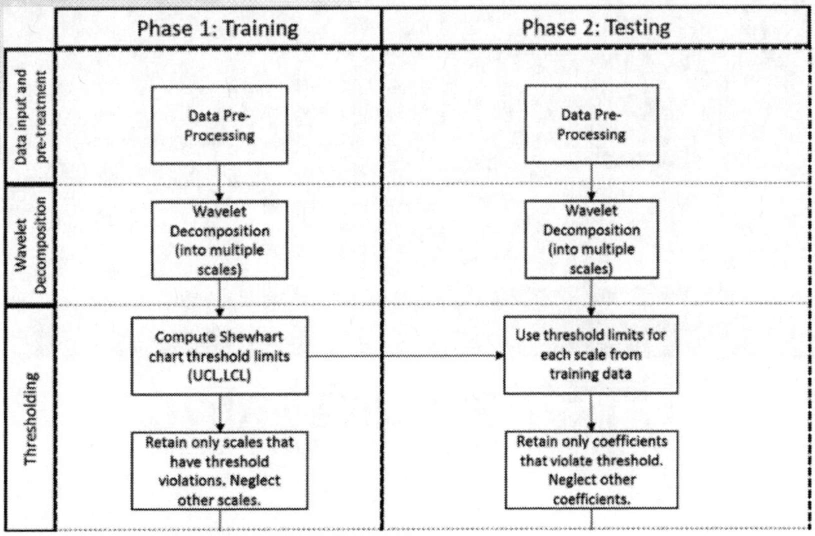

Fig. 7: Multiscale Shewhart Algorithm: Part I [2]

[2] M. Z. Sheriff, F. Harrou, and M. Nounou, "Univariate process monitoring using multiscale Shewhart charts," in 2014 International Conference on Control, Decision and Information Technologies (CoDIT). 2014, pp. 435–440.

14

Multiscale Shewhart Chart Algorithm

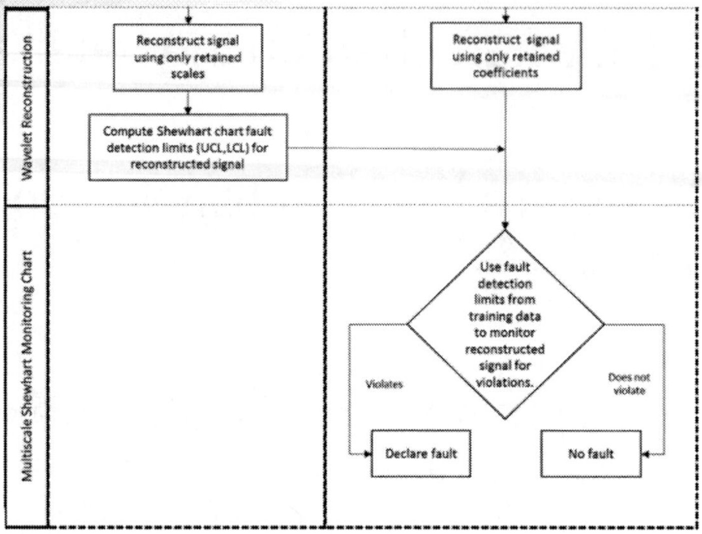

Fig. 8: Multiscale Shewhart Algorithm: Part II [2]

[2] M. Z. Sheriff, F. Harrou, and M. Nounou, "Univariate process monitoring using multiscale Shewhart charts," in 2014 International Conference on Control, Decision and Information Technologies (CoDIT), 2014, pp. 435–440.

15

Illustrative Examples

ENGINEERING
TEXAS A&M UNIVERSITY

Assumptions of the Shewhart Chart

1. Residuals contain a moderate level of noise
2. Residuals are independent (uncorrelated)
3. Fault-free residuals are normally distributed (Gaussian)

16

1. Dealing with High Noise Levels

- Data Generation
  - Standard Normal Distribution: zero mean, Gaussian
- Fault Generation:
  - Additive (Step) of ± 2 added at two locations
- Simulation repeated for noise standard deviation: σ=0.01 to σ=2 (Monte-Carlo: 5000 realizations)

17

1. Dealing with High Noise Levels

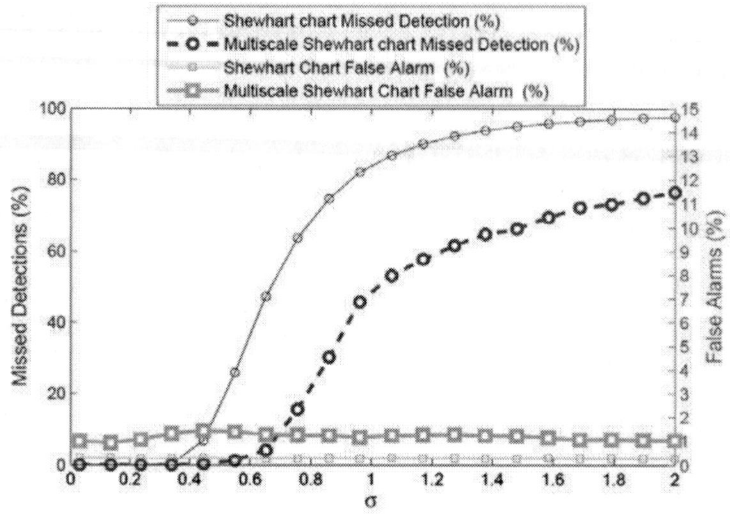

Fig. 9: Impact of High Noise Levels on performance (Monte-Carlo)

18

1. Dealing with High Noise Levels

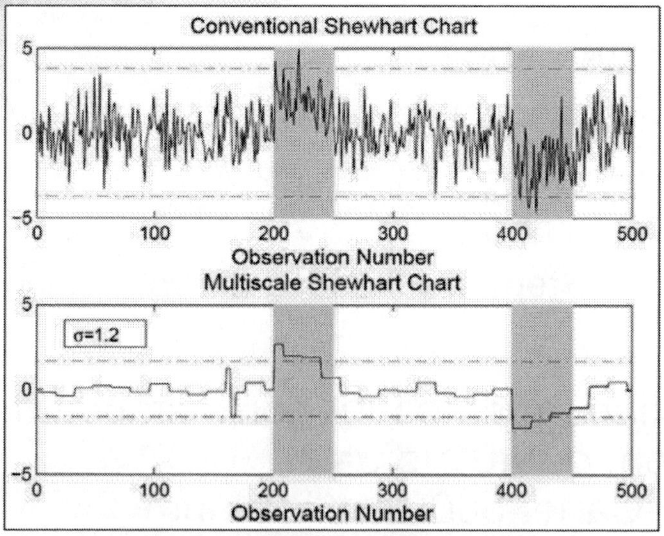

Fig. 10: Impact of High Noise Levels on performance

Gibbs Phenomenon: Production of artifacts during reconstruction

19

2. Dealing with Autocorrelation

- Data Generation
  - Autoregressive Model (AR(1)): $\tilde{z}_t = a_1 \tilde{z}_{t-1} + \xi_t$
    - $\tilde{z}_t = z_t - \mu$: deviation from targeted mean
    - ξ_t: standard normal noise
    - a_1: autoregressive coefficient

- Fault Generation:
  - Additive (step) of magnitude +3 added at one location

- Simulation repeated for autoregressive coefficient: a_1=0.3 to a_1=1 (Monte-Carlo: 5000 realizations)

20

2. Dealing with Autocorrelation

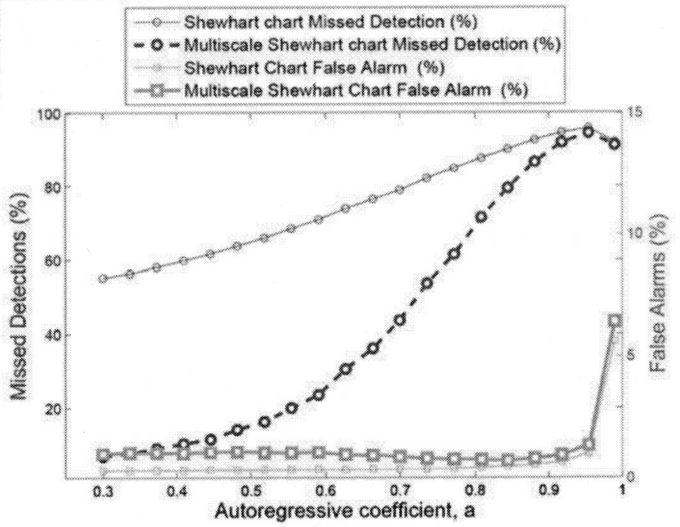

Fig. 11: Impact of Autocorrelation on performance (Monte-Carlo)

21

2. Dealing with Autocorrelation

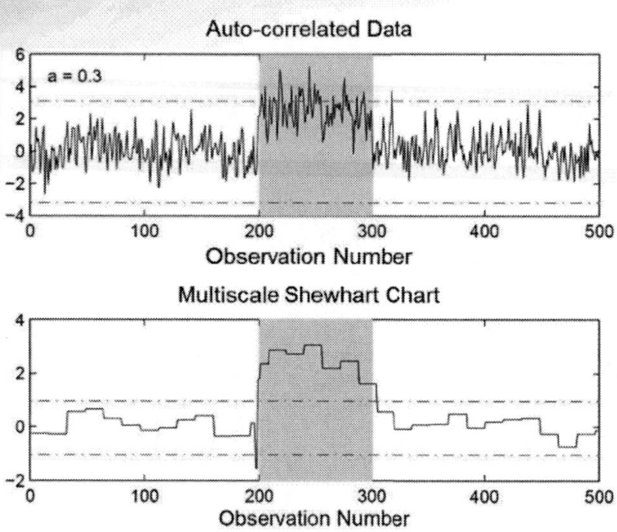

Fig. 12: Impact of Autocorrelation

Clearer detection with multiscale

22

3. Dealing with Deviation from Normality

- Data Generation
  - Chi-square distribution

$$\frac{1}{2^{\frac{k}{2}}\Gamma\left(\frac{k}{2}\right)} x^{\frac{k}{2}-1} e^{-\frac{x}{2}}$$

- Fault Generation
  - Additive step faults $\pm 3\sigma$ added at two locations (σ=std. dev. of data set)

- Simulation repeated for Shapiro-Wilk: SW=0.68 to SW=0.99 (Monte-Carlo: 5000 realizations)

23

TEXAS A&M UNIVERSITY | TEXAS A&M UNIVERSITY at QATAR

3. Dealing with Deviation from Normality

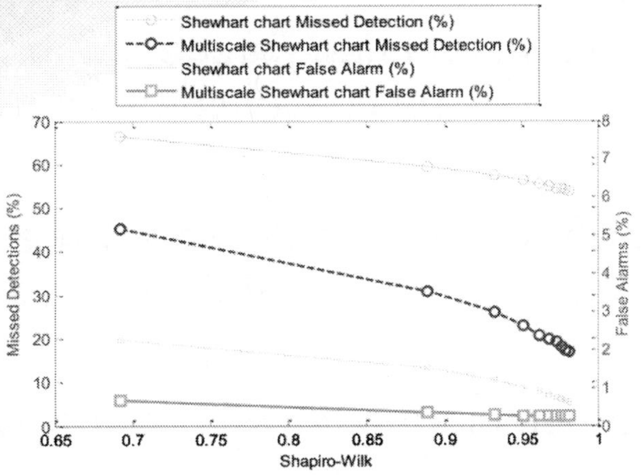

Fig. 13: Impact of Deviation from Normality on performance (Monte-Carlo)

Less Normal   More Normal

24

TEXAS A&M UNIVERSITY | TEXAS A&M UNIVERSITY at QATAR

3. Dealing with Deviation from Normality

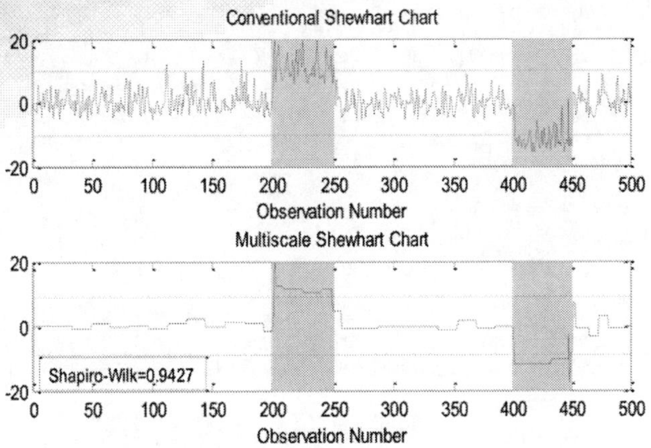

Fig. 14: Impact of Deviation from Normality on performance

Clearer detection and lower false alarm rates with multiscale

25

TEXAS A&M UNIVERSITY | TEXAS A&M UNIVERSITY at QATAR

Assessment Summary

- Multiscale Shewhart chart can deal with violations in all assumptions of the conventional Shewhart chart
- Multiscale Shewhart chart provides lower missed detection rates and comparable false alarm rates
- Advantages are brought forward primarily due to the nature of the detail signals

26

Application: Distillation Column

ENGINEERING
TEXAS A&M UNIVERSITY

Simulated Example: Distillation Column

- Purpose:
  - To monitor the composition of propane in the distillate feed of a distillation column
- Data Generation:
  - 1024 observations of the distillate feed (assumed noise-free)
  - Zero-mean Gaussian noise added
  - PLS model to obtain residuals
- Fault Generation:
  - Faults of $\pm 2.5\sigma$ and $\pm 3.5\sigma$ at two different locations

TEXAS A&M UNIVERSITY | TEXAS A&M UNIVERSITY at QATAR

Simulated Example: Distillation Column

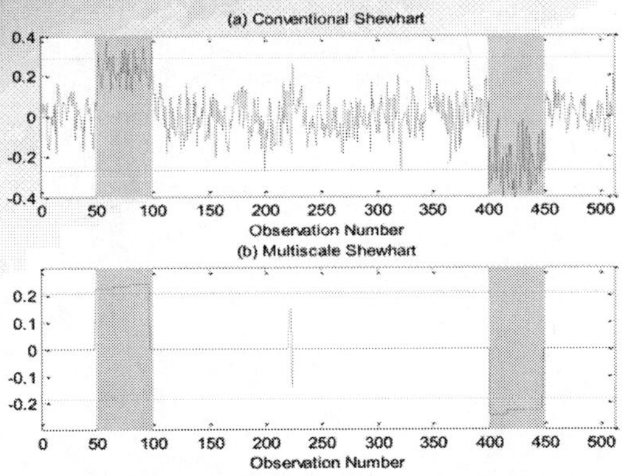

Fig. 15: Performance comparison for a step fault of 2.5σ

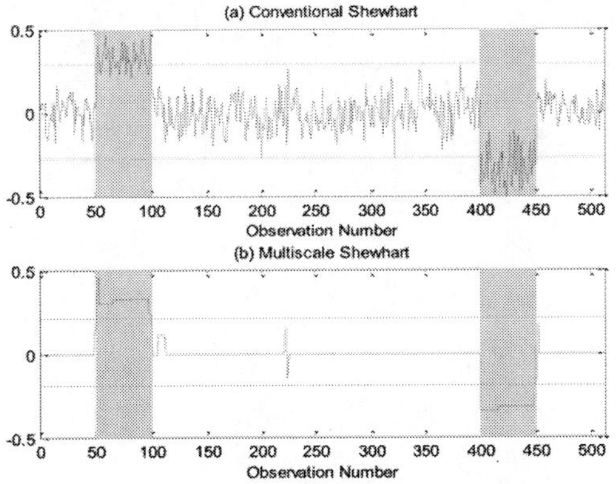

Fig. 16: Performance comparison for a step fault of 3.5σ

- Conventional chart can reasonably detect only large faults
- Multiscale chart is able to better detect both small and large faults

Conclusion

- ViolationViolation of assumptions cause a deterioration in performance of conventional Shewhart chart
- Multiscale representation has advantages that can help deal with violations in these assumptions
- When applied on the detail coefficients, the multiscale Shewhart chart provides better missed detection and false alarm rates for assumption violations

29

Thank You! Questions?

444955 Constrained Model Identification Using an Open-Equation Nonlinear Optimization Solver

Abstract:

In model identification, especially in industrial applications, one of the key steps is model refinement. This step is traditionally completed after the model is obtained through linear least squares estimation without constraints. After estimating the parameters of the model based on data, model refinement helps ensure that the model is appropriate for the desired goal. Model refinement involves such considerations: forcing material balance, forcing known steady-state gains and gain ratios, removing null models, forcing the exact co-linearity, and repairing ill-conditioned sub-matrices, consistency of steady-state models, and dynamic consistency relationships. As mentioned, these corrections are usually done manually. Proving a means to impose these constraints directly in the identification optimization problem instead of as a post-processing step has the potential to improve a variety of Model Predictive Control (MPC) applications. Applying prior knowledge and model consistency relationships leads to handling ill-conditioning for the purpose of better control performance [1, 2]. The other advantage of incorporating constraints in the identification step is to achieve smaller variance for parameter estimates possibly with less testing [1, 3].

In [2], Thorpe et al. incorporate different constraints such as dead time, settling time, and gains or gains rations for FIR or subspace models. There is a wide range of other possibilities for considering constraints for the identification procedure. Darby et al. in [1], summarized possible improvements in different steps of performing MPC. For example, consistency of steady-state models [4], and dynamic consistency relationships [5] can be imposed during the model identification step.

In this work, the goal of model identification is to achieve the minimum value for a function of error between the system response and the model while satisfying several constraints. The model should satisfy certain conditions so that it reflects known aspects of the system in order to perform well in closed-loop. In most traditional identification methods, these conditions are not imposed on the model identification problem. By imposing constraints, the model identification of Multi-Input Multi-Output (MIMO) systems along with discretization in time or space becomes a large-scale nonlinear constrained optimization problem.

Among different ways of formulating an optimization problem, the open-equation format has found more attention recently. Writing the equations in residual form allows the differential term to be expressed in implicit form [6]. This strategy has the ability to solve problems with a large number of variables and equations simultaneously without nested convergence loops [7,8]. Although the open-equation method still has some challenges to solve very large-scale problems, it has the potential to be efficient for constrained model identification. This work investigates the effect of imposing different constraints on the model identification step, and shows the capabilities and limitations of the open-equation nonlinear solver to solve the estimation problem.

References:

1. Darby, M. L., Nikolaou, M., MPC: Current practice and challenges, Control Engineering Practice, 2012, 20:328-342.

2. Thorpe, P. J., Nicholson, D, Osta, S., Harmse M., Constrained Model Identification: Use Knowledge of the process to get better models from less data, AIChE, 2010.

3. Soderstrom, T., and Stoica, P., System identification, Prentice Hall, 1989.

4. Skogestad, S., Consistency of Steady-State Models Using Insight about Extensive Variables, Ind. Eng. Chem. Res. 1991, 30:654-661.

5. Eskinat, E., Dynamic consistency relations for process modeling. AIChE Journal, 2003, 49(8): 2224–2227.

6. Hedengren, J. D., Asgharzadeh Shishavan, R., Powell, K. M., and Edgar, T. F., Nonlinear Modeling, Estimation and Predictive Control in APMonitor, Computers & Chemical Engineering, 2014, 70:133-148.

7. AspenTech, Getting Started Using Equation Oriented Modelling, 2011.

8. Meixell, M. D., Gochenour, B. and Chen, C., Industrial Applications of Plant-Wide Equation-Oriented Process Modeling, Advances in Chemical Engineering, 2011, 40: 119–152.

AUTHOR INDEX

AUTHOR INDEX

AIChE
120 Wall Street, FL 23
New York, NY 10005-4020

ISBN 978-1-5108-2494-2

9 781510 824942

CURRAN ASSOCIATES INC.
proceedings.com

Poster Sessions 2022

Held at the 2022 AIChE Annual Meeting

Phoenix, Arizona, USA
13-18 November 2022

AIChE
The Global Home of Chemical Engineers